CHEMICAL ENGINEERING METHODS AND TECHNOLOGY

TRACE ELEMENTS

ENVIRONMENTAL SOURCES, GEOCHEMISTRY AND HUMAN HEALTH

CHEMICAL ENGINEERING METHODS AND TECHNOLOGY

Additional books in this series can be found on Nova's website under the Series tab.

Additional e-books in this series can be found on Nova's website under the e-book tab.

BIOCHEMISTRY RESEARCH TRENDS

Additional books in this series can be found on Nova's website under the Series tab.

Additional e-books in this series can be found on Nova's website under the e-book tab.

CHEMICAL ENGINEERING METHODS AND TECHNOLOGY

TRACE ELEMENTS

ENVIRONMENTAL SOURCES, GEOCHEMISTRY AND HUMAN HEALTH

DIEGO ALEJANDRO DE LEON
AND
PALOMA RAQUEL ARAGON
EDITORS

nova publishers

Nova Science Publishers, Inc.
New York

For permission to use material from this book please contact us:
Telephone 631-231-7269; Fax 631-231-8175
Web Site: http://www.novapublishers.com

LIBRARY OF CONGRESS CATALOGING-IN-PUBLICATION DATA

Trace elements : environmental sources, geochemistry, and human health / [edited by] Diego Alejandro De Leon and Paloma Raquel Aragon.
p. cm.
Includes index.
ISBN 978-1-62081-376-8 (hardcover)
1. Trace elements. 2. Trace elements--Environmental aspects. I. De Leon, Diego Alejandro. II. Aragon, Paloma Raquel.
QD139.T7T726 2011
572'.515--dc23

2012008875

Published by Nova Science Publishers, Inc. † New York

CONTENTS

PREFACE

This book examines the environmental sources, geochemistry and human effects of trace elements. Topics discussed include an assessment of cancer and non-cancer health risk caused by trace element content in Belgrade; flow opto-sensing applied to the analysis of trace elements; trace metals in fruits and vegetables and their effect on human health; excessive livestock mineral supplementation on environmental pollution; assessing the geochemistry of trace elements in canal sediments; and trace element deficiencies in crops.

Chapter 1 – The urban atmosphere is subjected to large inputs of anthropogenic contaminants produced by both stationary (power plants, industry and residential heating) and mobile sources (related to traffic). Trace elements are widely dispersed in the environment and can have toxic effects on the biosphere and human health. The characterization of airborne trace elements is relevant to identification of variability and sources of atmospheric pollutants. The transport of trace elements has also attracted much attention. Extensively employed collection of atmospheric deposition using bulk sampling devices, or using air samplers of suspended particulate matter (PM) offer practical approaches for monitoring of atmospheric trace elements in the environment. However, the instrumental sampling methods are frequently limited by the high cost and difficulties in carrying out extensive monitoring in time and space. There has been increasing interest in the use of indirect monitoring methods, such as biomonitoring. Thus, mosses have been recognized as valuable airborne trace element biomonitors. In urban areas, where mosses are often scarce or even absent, the active biomonitoring (moss bags) has been developed with the aim of spatial and/or temporal assessment of atmospheric deposition. Since 2003 several studies have been carried out in the urban area of Belgrade using both instrumental and biomonitoring methods_with the primary objective of assessing anthropogenic impacts on the environment by determination of trace element atmospheric pollution. Suspended PM_{10} were collected on filters using air sampler, while monthly atmospheric bulk depositions were collected monthly using bulk collectors. Active biomonitoring of trace elements with moss bags was performed with the aim to test the suitability of the moss, *Sphagnum girgensohnii* Russow, as an alternative method for air quality studies in urban areas. The elemental composition (Al, V, Cr, Mn, Fe, Ni, Cu, Zn, Cd, and Pb) of the moss samples was analyzed using instrumental neutron activation analysis (INAA), atomic absorption spectrometry (AAS), and inductively coupled mass spectrometry (ICP-MS). For the source apportionment purpose Unmix receptor model was used and the main source profiles were identified. Effects of regional sources were investigated using hybrid receptor models for identification of trace elements source regions: Potential Source

Contribution Function (PSCF) - describes the spatial distribution of probable geographical source locations inferred by using trajectories arriving at the sampling site; and Concentration Weighted Trajectory (CWT) model - determines the relative significance of potential sources. The impact of trace elements in PM_{10} on air quality and human health is considered as significant at receptor areas. PM_{10} penetrate into the human respiratory system, accumulate in human body and can exert health problems. The assessment of cancer and non-cancer health risk caused by the trace element content of PM_{10} in Belgrade was conducted using the US EPA health risk assessment model.

Chapter 2 – The determination of trace amounts of inorganic ions is of growing interest in several fields including environmental analysis, process control, biology, medicine, etc., and there is now a particular need for simple and fast field analytical tests. In contrast to well-established laboratory methods, such as atomic absorption or emission spectrometry, mass spectrometry or voltammetry, chemical sensors have attracted a great deal of interest in the last decade because of their great potential for in the field, *in situ*, continuous and remote, if required, applications. Flow-through optosensing emerged from the combination of Solid-Phase Spectroscopy (SPS) and Flow Injection Analysis (FIA). In SPS, an active solid support is used to pre-concentrate the target analyte and its spectroscopic detection is performed on the same solid support microbeads. The most remarkable features of SPS are its high sensitivity (it can be easily increased one hundred-fold due to the concentration of the target analyte on the solid beads) and selectivity (the interaction between the target chemical species and the solid phase beads is usually different from that one of co-existing species). In addition, the implementation of SPS in flow methods of analysis adds the inherent advantages of flow analysis, improving the analytical methods in terms of rapidity, commodity, automation, less consumption of reagents, etc. In these flow optosensors, the (micro) zone of the solid phase, where the signal is continuously monitored, is surrounded by a continuous stream flowing through it. The separation and retention of the species of interest on the solid phase takes place in the detection area itself and the spectroscopic determination is simultaneously performed. In this chapter, the authors will first describe the fundamentals of SPS from its origin, including some relevant examples. After that, they will focus on the versatility and potential of flow-through optosensing for the analysis of trace elements in a wide variety of samples related to human health, such as clinical samples (drugs, urine, and serum), food or environmental waters, emphasizing the most recent advances and future trends in this field.

Chapter 3 – Trace metals occur naturally in the earth's crust, but human activities have greatly increased the mobilization and bioavailability of metals and, therefore, the chance of exposure to harmful concentrations. In fact, they along with other pollutants are discharged in the environment through industrial activity, car exhaust, heavy-duty electric power generators, refuse burning and use of pesticides in agriculture etc. Natural baseline levels of trace metals in air, soil, rivers, lakes, and oceans are usually low, but in certain forms, and at sufficiently high concentrations, trace metals can be toxic to organisms. Trace elements play an important role in biological and biochemical reactions in the living cells. They are stable and cannot be metabolized by the organisms, so, through the food, they tend to bio-accumulate. If the accumulation rate is higher than the detoxification rate, they can reach toxic concentration levels. Toxic metals in the human body can "replacing" essential nutrients, especially if deficient; for example Pb can replace Ca, Al can replace Mg, etc.. The substitution allows the enzyme systems to continue its function, but, if the intoxication degree exceed the tolerable limits, serious diseases take over. The major source of trace metals for man is the food, in particular, fruit and vegetables that are essential components of the diet, by contributing protein, vitamins, iron, calcium and other nutrients in short supply. In recent years their consumption is gradually increasing, particularly in the urban community. However, because of the increased_environmental pollution, fruit and vegetables can also contain trace metals over a wide range of concentrations. That because the plants take off them from soil as well as from air particulate. In addition the contamination of fruit and vegetable products may be due also through irrigation with contaminated water, the use of fertilizers and metal-based pesticides, industrial emissions, transportation, the harvesting process, storage and/or at the point of sale. To monitor metal contents in these products is very important for crop yield technology, food nutrition and health impact. Many studies show that the trace metals at relatively high concentration may disturb the normal physiology (the normal functions of central nervous system, liver, lungs, heart, kidney and brain, produce hypertension, abdominal pain, skin eruptions, intestinal ulcer) until to produce different types of cancer.

Chapter 4 – Trace elements are essential to maintain physiological functions and it is well demonstrated that certain mineral supplementation in livestock can improve productivity. Mineral supplements are routinely incorporated into concentrate feed, even when not necessary, at doses largely (up to 10-fold) exceeding the physiological requirements; this is possible because there are fairly wide "safety margins" to avoid mineral toxicity, even though toxic effects on animals can appear when given at supra-optimal concentrations. A significant proportion of minerals that are given to livestock are excreted into urine and faeces, and this proportion increases as the margin of mineral supply over mineral requirement increases. This chapter addresses the main consequences of excessive livestock mineral supplementation on environmental pollution, the toxic effects on susceptible species when exposed to mineral enriched-slurries, and the pros and cons for human health of consuming animal products with a high mineral content.

Chapter 5 – Although trace elements are natural constituents of the environment, several of them present high toxicity, long persistency and quick accumulation in living organisms. For these reasons identification of trace element contamination sources and evaluation of their environmental impact are essential. Trace element pollution in estuaries has been considered a wide world problem for a long time. The principal pollution sources are atmospheric deposition (volcanic activity, combustion of fossil fuels, etc), land run-off (agricultural, mines, etc), direct industrial and urban inputs, fisheries and recreational uses. Once in water, trace elements are mainly adsorbed on suspended particle matter or colloids and deposited by flocculation in sediments, where they present different chemical forms. Thus sediments are important deposits of trace elements and hence an important tool to assess metal pollution in estuaries. Bibliographic data on the concentration of trace elements (As, Cd, Cr, Cu, Fe, Mn, Ni, Pb and Zn) measured in sediments from different estuaries of the world has been collected, summarized and compared with those experimentally obtained in two of the most important estuaries in the Basque Country (Urdaibai and Nerbioi-Ibaizabal). In general, it can be said that the concentration of trace element in sediments from the Nerbioi-Ibaizabal and Urdaibai estuaries are not alarming in comparison with other estuaries. Different tools have been used to identify diffuse and point sources in both estuaries.

Chapter 6 – The contamination of sediments by potentially harmful trace elements is of major concern, because of their toxicity, persistence and bioaccumulative nature. In many industrial and urban areas, sediments are the largest repository for, and potential source of these elements. Assessing trace element contamination of sediment is complicated since these elements are a ubiquitous, naturally occurring component of sediment, their concentrations in un-contaminated sediment can vary by orders of magnitude over relatively small spatial scales, and naturally occurring and anthropogenically introduced metals tend to accumulate in the same areas. Due to their particle reactivity, trace elements tend to accumulate in sediments and may persist in the environment long after their primary source has been removed. Element adsorption increases with decreasing grain size of the sediment. Thus, the metal concentrations significantly enriched in fine-grained sediment rich in clay minerals. Because of that, in geochemical and ecochemical studies, the normalization of elemental content using an immobile element is common practice for correcting grain-size effects and dilution by sedimentary phases such as carbonates and silica. The aims of the presented study were: (1) to assess the geochemistry of trace elements (Cr, Zn, Cd, Ni, Pb, and Cu) in canal sediments (Danube alluvial formation) and Tisa River sediments and its tributaries, after normalization of the elements data to a conservative element; (2) to discriminate natural and anthropogenic contributions; and (3) to identify possible sources of pollution. As the element for sediment normalization was used Al, since it represents aluminosilicates, the main group of minerals generally found in the fine sediment fractions. Flame atomic absorption spectrometry (FAAS) was used to determine the levels of the investigated elements. Multivariate statistical methods have been applied in this research to identify similarities in the sampling stations and to assess the relationship between the elements. The results from this study show that investigated sediments are affected by combined trace elements contamination. Also, there was shown a significance of applying geochemical normalization and statistical analysis for assessing trace element contamination and for distinguishing between natural and anthropogenic sources of trace elements in the sediments.

Chapter 7 – Zinc (Zn) deficiencies occur all over the world. Due to the naturally high soil fertility of the Pampas region, Argentina, macro and micronutrients have not been applied to soils through fertilization in a massive way. As a consequence, some trace elements deficiencies were noticed in some forage crops during the last years as a result of nutrient depletion. Trace mineral deficiencies were detected in dairy cows, causing a variety of suboptimal performances such as poor production and reproductive inefficiencies. Zinc and copper are the trace elements present in largest amounts in Buenos Aires City's biosolids. Soils amended with biosolids have been reported to have high Zn availability. Therefore, biosolids land application may raise the concentration of Zn in forage crops, reducing the need for mineral fertilisers or supplements in the diet of dairy farms. This chapter address sound soil and crop management practices that may not only enhance crop productivity but also help in reducing Zn deficiencies in dairy cattle.

In: Trace Elements ISBN: 978-1-62081-376-8
Editors: Diego A. De Leon and Paloma Raquel Aragon

Chapter 1

A STUDY OF AIRBORNE TRACE ELEMENTS IN BELGRADE URBAN AREA: INSTRUMENTAL AND ACTIVE BIOMONITORING APPROACH

M. Aničić, Z. Mijić, M. Kuzmanoski, A. Stojić, M. Tomašević, S. Rajšić and M. Tasić
Institute of Physics, University of Belgrade, Serbia

ABSTRACT

The urban atmosphere is subjected to large inputs of anthropogenic contaminants produced by both stationary (power plants, industry and residential heating) and mobile sources (related to traffic). Trace elements are widely dispersed in the environment and can have toxic effects on the biosphere and human health. The characterization of airborne trace elements is relevant to identification of variability and sources of atmospheric pollutants. The transport of trace elements has also attracted much attention. Extensively employed collection of atmospheric deposition using bulk sampling devices, or using air samplers of suspended particulate matter (PM) offer practical approaches for monitoring of atmospheric trace elements in the environment. However, the instrumental sampling methods are frequently limited by the high cost and difficulties in carrying out extensive monitoring in time and space. There has been increasing interest in the use of indirect monitoring methods, such as biomonitoring. Thus, mosses have been recognized as valuable airborne trace element biomonitors. In urban areas, where mosses are often scarce or even absent, the active biomonitoring (moss bags) has been developed with the aim of spatial and/or temporal assessment of atmospheric deposition.

Since 2003 several studies have been carried out in the urban area of Belgrade using both instrumental and biomonitoring methods_with the primary objective of assessing anthropogenic impacts on the environment by determination of trace element atmospheric pollution. Suspended PM_{10} were collected on filters using air sampler, while monthly atmospheric bulk depositions were collected monthly using bulk collectors. Active biomonitoring of trace elements with moss bags was performed with the aim to test the suitability of the moss, *Sphagnum girgensohnii* Russow, as an alternative method for air quality studies in urban areas. The elemental composition (Al, V, Cr, Mn, Fe, Ni, Cu, Zn, Cd, and Pb) of the moss samples was analyzed using instrumental neutron activation analysis (INAA), atomic absorption spectrometry (AAS), and inductively coupled mass spectrometry (ICP-MS). For the source apportionment purpose Unmix receptor model was used and the main source profiles were identified. Effects of regional sources were investigated using hybrid receptor models for identification of trace elements source regions: Potential Source Contribution Function (PSCF) - describes the spatial distribution of probable geographical source locations inferred by using trajectories arriving at the sampling site; and Concentration Weighted Trajectory (CWT) model - determines the relative significance of potential sources. The impact of trace elements in PM_{10} on air quality and human health is considered as significant at receptor areas. PM_{10} penetrate into the human respiratory system, accumulate in human body and can exert health problems. The assessment of cancer and non-cancer health risk caused by the trace element content of PM_{10} in Belgrade was conducted using the US EPA health risk assessment model.

1. INTRODUCTION

Air pollution is a serious problem in many parts of the world, affecting people and the economy of countries, leading to forest decline, loss in agricultural production and diminished status of the population. A wide array of substances including greenhouse gases, organic dust, and particulate matter is being emitted from natural (meteoric, terrestrial, marine, volcanic, erosion and surface winds, forest fires, biogenic) and anthropogenic (fossil fuel combustion, industry, automobiles, agricultural) sources. The main global concern is on the climate effects of greenhouse gases. However, airborne particulate matter (PM), as well as PM bound trace elements, is more directly affecting human health, leading to the need for strict governmental regulations on air particle and trace element concentrations. Namely, after emission, the trace elements are subjected to physical, chemical, and photochemical transformations, which ultimately decide their fate and atmospheric concentrations. Like other air pollutants, trace elements do not remain confined near the source emission, but spread over distances, depending upon topography and meteorological conditions, especially wind direction and speed, and vertical and horizontal thermal gradients, transcending natural and political boundaries [1,2,3].

Although some elements (or their compounds) can occur in vapor states (Hg), almost all atmospheric trace elements have been detected in aerosols. More than 40 trace elements are measured in atmospheric particulate matter samples [4,5]. Depending on their sources, these elements can be found in particles of different diameters between 0.01 μm and 100 μm. Large particles have a tendency to precipitate quickly near the discharge sources, but particles of sizes less than 10 μm and low density exhibit long term existence, from days to months, and may be transported to distant regions. There is evidence that trace elements of anthropogenic sources are widely distributed in the atmosphere and occur even at high altitudes in mountain regions and in the Arctic [6].

Since 1979 the Convention on Long-Range Transboundary Air Pollution (CLRTAP) has addressed some of the major environmental problems of the UNECE region through scientific collaboration and policy negotiation. The Convention has been extended by eight protocols that identify specific measures to be taken by Parties to cut their emissions of air pollutants.

The trace element concentrations in the atmosphere have been monitored in some countries for over 30 years. Also, retrospective studies of trace element enrichment in ice layers, ombrotrophic peatlands and tree-rings show a steady increase of several trace metal (Pb, Cd, etc.) contents due to anthropogenic emissions. However, some recent observations have shown decreasing trends for several metals, e.g. Fe, Pb, Cr, Cd, Hg, As, and Be. A significant reduction of the Pb content in airborne particulates has recently been reported for some urban sites due to replacement of leaded with unleaded gasoline [6].

The transport and mobilization of trace elements have also attracted much attention [7,8] and regional and seasonal differences in the element distribution are evident. The elemental composition of airborne particulates at different locations may vary over almost tree orders of magnitude indicating a strong effect of local sources. The highest element concentrations are usually reported for urban and industrial areas.

The urban atmosphere is subjected to large inputs of anthropogenic contaminants produced by both stationary sources and mobile sources related to traffic. So, in urban environment atmospheric concentrations of several trace elements are typically 5-10 times higher than those from rural sites [1,5]. In general, elements such as Pb, Fe, and Cu have the highest concentration, while elements such as Co, Hg, and Sb are characterized by low concentration. These elements arise from dozens of different sources including combustion of coal, oil, wood burning, steel furnaces, boilers, smelter, dust, waste incineration, and break wear [2]. Furthermore, atmospheric concentration of elements from anthropogenic sources (e.g. Pb, Br, V, Zn, etc.) can show significant seasonal variation. Winter maximum is frequently observed and usually attributed to increased fuel combustion and the higher frequency of atmospheric temperature inversion [5].

Trace elements are widely dispersed in the environment and their interactions with different natural components result in toxic effects on the biosphere. Air pollutants have a significant, direct and indirect, impact on human health. The size of particles, especially PM_{10} (aerodynamic diameter less than 10 μm) that are easily inhaled, are of a great significance. In addition, they may be absorbed by aerial parts of vegetables and thus included into the food chain [6].

Occupational exposure to PM_{10} is of particular concern and should be monitored because clean air is considered a basic requirement for humans. The World Health Organization (WHO, 2005) has updated the air quality guidelines for Europe. They mentioned that the "Guidelines values" are not just numbers, but they incorporate recommendations that will help to reduce harmful effects of organic and inorganic particulates to humans and vegetables. However, there is limited information on guideline values for inorganic elements [6].

Most trace elements in terrestrial ecosystems originate from atmospheric wet and dry deposition [5]. The chemical composition of rainwater is closely associated with air pollution and greatly contributes to the deposition of pollutants [6]. From a biogeochemical perspective, the characterization of atmospheric deposition is relevant in order to identify the variability and sources of the atmospheric pollutants [9]. Within the framework of the Cooperative Programme for Monitoring and Evaluation of the Long-range Transmission of Air Pollutants in Europe (EMEP) measurements of PM and trace metals, as highly toxic species, have been introduced.

The extensively employed collection of atmospheric deposition using bulk sampling devices or air samplers of particulate matter, offer practical approaches for monitoring of atmospheric trace elements in the environment [9,10]. Studies of spatial and temporal variation of atmospheric aerosol particles also gained significance and resulted in an increased interest in the use of analytical techniques capable to measure the size, morphology and chemical composition of individual aerosol particles. Such data are essential for an understanding of particle formation, transport, transformation and deposition mechanisms, as well as the impact of particles inhaled into respiratory system.

The majority of reliable instruments generally make use of a validated sampling head to select the PM fraction of the ambient airborne particles. The collected particles are analysed in two main ways, gravimetric and direct-reading monitors. The first one is related to cumulative samplers in which the PM are deposited on a filter over a sampling period of 24 hours. The mass of particles collected on the filters is determined by weighing. Direct-reading monitors select PM which are either deposited on a filter, with continuous assessment of the change of a property of the filter due to their presence, or passed through an optical sensing region. The reference methods, based on the manual weighing of filters, are clearly unsuited to real-time automated data production that can be disseminated to the public in the same way as for gaseous pollutants. Commonly used automatic methods across Europe include Tapered Element Oscillating Microbalances (TEOM), which measure the increasing mass of a filter by its effect on the resonant frequency of a vibrating support; beta attenuation monitors, which measure the increasing amount of material on a filter by its absorption of electrons emitted by a weak beta-source; and optical monitors, which can gauge the size of individual particles from signals scattered from a light beam and integrate this into a total volume of particles.

The complexity of aerosol processes in environment is so great that it leads to large uncertainties in understanding of their role in many important environmental issues. A single type of observation platform can not, by itself, provide sufficient data to fulfil our current observational needs. A combination of surface-based (in situ and remote sensing), suborbital (aircraft and balloons), and satellite observations are needed.

For monitoring of the air quality at street level, i.e. where people are in direct contact with aerosols, in situ measurements are most adequate. PM sampling is performed for the variety of purposes, including mass concentration determination for compliance with standards. Chemical analysis of filter deposits can not be separated from the methods used to obtain the sample. Before the choice of equipment is made permanent, it is essential that these methods are evaluated and optimized. Techniques commonly used for trace element analysis of environmental samples include atomic absorption spectrometry (AAS), graphite furnace atomic absorption spectrometry (GF-AAS), inductively coupled plasma atomic emission spectrometry (ICP-AES), and inductively coupled plasma mass spectrometry (ICP-MS). New developments in XRF analysis can provide analysts with sensitivities that approach and match those of GF-AAS and ICP-MS.

PM_{10} enter the respiratory system through inhalation and can be accumulated in the upper respiratory tract, while smaller particles can penetrate deeper into the respiratory system, causing health problems. Furthermore, these particles often contain toxic trace elements. Studies showed that chronic exposure to PM_{10} results in adverse health effects [11], increasing the risk of developing cardiovascular and respiratory diseases and lung cancer. Among PM_{10}-bound trace elements, of particular concern are As, Cd, Cr, Ni, and Pb, which are human carcinogens. US EPA has classified As, hexavalent Cr and Ni (nickel refinery dust and nickel subsulfide) as group A, known human carcinogens, concluding that there is enough evidence that they cause cancer in humans by inhalation. Cd and Pb are classified as group B - probable human carcinogens, due to limited or inadequate evidence that they can cause cancer in humans based on epidemiological and animal studies. These and other elements can also cause non-carcinogenic health problems. Analysis of elemental content of

PM_{10} allows assessment of health risk (both carcinogenic and non-carcinogenic) associated with exposure to PM_{10}-bound trace elements through inhalation.

The instrumental sampling methods are frequently limited by the high cost and difficulties in carrying out extensive monitoring in time and space. Thus, there has been increasing interest in the use of indirect monitoring methods, such as the use of organisms that act as bioaccumulators. Biomonitoring is a rapid and economical method that has commonly been used for assessing environmental quality and potentially detrimental effects of pollutants to the biosphere [12,13]. Various effective bioindicators have been used so far for assessment of the state of natural ecosystems. These include mosses and lichens that are commonly regarded as the best bioindicators of air quality, as they can accumulate elements to a far greater level than it is necessary for their physiological needs. However, leaves or bark of various tree species, both evergreen and deciduous, have also been used for biomonitoring of air pollution [5,14,15,16,17,18,19].

Many studies have demonstrated the ability of moss to absorb and accumulate atmospheric pollutants in tissue. Root and cuticule absence makes them find their nutritive elements in wet and dry atmospheric deposition [20]. Mosses have also been recognized as valuable biomonitors in the assessment of temporal trends in trace element accumulation [21], and in spatial variations across national boundaries [22]. Moreover, mosses can accumulate and concentrate toxic substances that may be present even in low concentrations in the local environment. Uptake and retention of elements by moss is aided by: *(i)* numerous small leaves and intricate surfaces (large area/volume ratio); *(ii)* high permeability of tissue to water and elements; *(iii)* high water retention capacity; *(iv)* high cation exchange capacity due to binding sites on the cell wall [5,23]. Generally, accumulation of trace elements depends on their abundance in air, their solubility in water, water availability and humidity [24,25]. Trace elements may be deposited onto the moss surface either as dry particulates or as a material dissolved and/or suspended in precipitation. The elements may be retained by particulate entrapment, physicochemical processes such as ion exchange, as well as by the passive and active intracellular uptake [26].

The use of native mosses as biomonitors is a convenient way of determining levels of trace elements atmospheric deposition [20,21,26,27,28]. However, if epiphytic mosses have been difficult to find at locations of interest, such as in urban and industrial areas, transplanted mosses have been employed as an option. The "moss bags technique" [27,29,30,31,32,33,34] is one of the active biomonitoring methods, where suitable moss species are sampled from an area with negligible influence from air pollution, properly cleaned from foreign materials, packed into nylon mesh bags, and then exposed at specific locations for defined periods of time to trap deposited elements.

The *Sphagnum* moss species are especially recommended as the most suitable moss for active biomonitoring of trace elements due to several features including a large surface area and number of protonated anionic functional groups (ion exchange sites) in the form of uronic acids (>50% cation exchange capacity - CEC), while phenolic compounds are responsible for about 25% of CEC) [35,36,37]. Moss of the genus *Sphagnum* are most suited for the moss bag method due to their very high element retention properties [38]. Mäkinen [39] found that bag filled with peat and cotton wool has retention capacities of only 43 and 35% respectively, in comparison with the capacity of a typical *Sphagnum* moss bag.

The exposure period is especially critical in active moss biomonitoring surveys [5]. If exposure time is too long, saturation of exchange sites on the moss membrane may occur, and preferential displacement or exchange of elements as well. Also, some elements may "occupy" the exchange sites on moss membrane cells, so that they are effectively immobilized [36]. It was found by Little [30] and Ratcliffe [40] that an exposure period of four weeks is optimal. Goodman *et al.* [41] found that the retention of metals by moss-bags was directly proportional to the exposure time and was linear for up to ten weeks for Co, Ni, Cu, Zn, Cd, and Pb. Displacement and saturation effects can be avoided by using the minimum exposure periods necessary for detecting trends. Another important factor is the bag design, which should keep the humidity of the sample at a stable level to avoid drying [13].

Since 2003, instrumental monitoring and active moss biomonioring of airborne trace elements have been conducted in the urban area of Belgrade in the consecutive experiments and a review of the results is given in this chapter.

2. Study Area

The investigations presented in this chapter were conducted in Belgrade (φ = 44° 49' N, λ = 20° 27` E, H_s = 117 m), the capital of the Republic of Serbia, with about 1.6 million inhabitants (Figure 1).

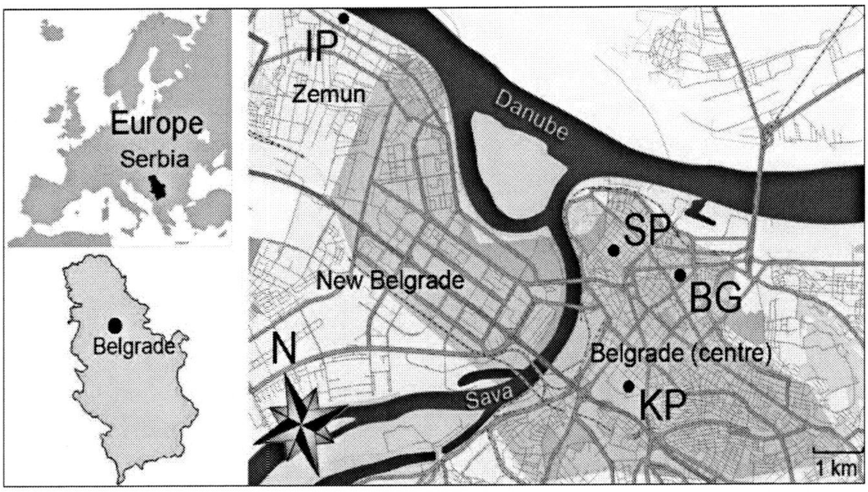

Figure 1. City of Belgrade and the Surrounding Area.

The city is situated at the confluence of the rivers Sava and Danube in the inner part of Balkan Peninsula. It is characterized with moderate continental climate with cold winters and warm summers. The annual average temperature is 12°C, the hottest month is July, with an average temperature of 22°C; annual precipitation is about 700 millimeters.

Significant sources of ambient trace elements in Belgrade are old vehicles (in average, passenger cars are more than 10 years old). The total number of vehicles is about 500,000 including heavy-duty lorries and over 1,000 city buses using diesel. This could be asignificant source of ambient PM_{10}, as diesel vehicles emit 10-20 times more particulate matter than gasoline vehicles [42]. At the time of the study discussed here, leaded petrol was widely used in Belgrade. In addition, there are several large heating plants in Belgrade, run with natural gas or crude oil and many smaller plants run only with crude oil. Fuel used for domestic heating is mainly coal or crude oil, as well as natural gas introduced during the last few years. In the winter, severe air pollution in the form of aerosol smog occurs frequently in the urban area of Belgrade, particularly under meteorologically calm (wind speed < 2 m s^{-1}) and stable conditions.

At several heavy traffic sites in Belgrade during the last decades instrumental monitoring and biomonitoring of trace elements has been conducted. The map of the urban Belgrade area with the studied sites (SP, BG, KP, and IP) is presented in Figure 1.

3. TRACE ELEMENTS IN PM_{10}

Identification of local sources in Belgrade area has been recognized as important in order to obtain a better understanding of the ambient air quality and to develop emission reduction strategy [43]. Chemical characterization of suspended particles, as well as application of different models, is necessary to apportion PM ambient concentration to their sources [44]. In order to identify sources, multivariate receptor modeling can be applied to the observed PM_{10} element composition data. As the strong negative health effect of exposure to the inhalable particulate matter and trace element content in the urban environment has been confirmed, the related study in Belgrade has been conducted since 2003. The obtained results, as well as brief description of methodology and receptor models used, are given below.

3.1. Sampling, Analysis and Modeling

Suspended PM_{10} particles were collected on preconditioned and pre-weighed Pure Teflon and Teflon-coated Quartz filters (Whatman, 47 mm diameter, 2 μm pore size) using MiniVol air sampler provided with PM_{10} cutoff inlets. The elemental composition (Al, V, Cr, Mn, Fe, Ni, Cu, Zn, Cd, and Pb) of the PM_{10} samples was determined by the atomic absorption spectroscopy (AAS). The sampling methodology was described in detail elsewhere [45,46,47]. Emission sources have characteristic chemical and/or physical patterns, and it may be useful to examine the relationship between specific source emissions and health effects. These source composition or physical properties profiles permit the contributions of these sources to the airborne PM mass to be apportioned. Therefore, more targeted control strategies could be devised to focus on sources that contribute significantly to the trace elements content of PM and related health problems. In the field of atmospheric sciences receptor models aim to reconstruct the impacts of emission from different sources of atmospheric pollutants based on ambient data measured at the monitoring sites [48,49,50]. These methods are called receptor-oriented or receptor models since they are focused on the behavior of the ambient environment at the point of impact, as opposed to the source-oriented dispersion models that focus on the transport, dilution and transformations that begin at the source and continue until the pollutants reach the sampling or receptor site. One of the most commonly used models is Unmix and the latest version of Unmix software is available from the US Environmental Protection Agency [51]. The concepts underlying Unmix have already been presented in geometrical and intuitive manner [52] and mathematical details are presented elsewhere [53]. There are many published papers related to ability of Unmix to reconstruct emission sources for various pollutants [49,54,55]. In this study Unmix model has been used to analyze the three-year (July 2003 to December 2006) PM_{10} data sets, collected in Belgrade, for source apportionment purpose.

To estimate the local source impacts from various wind directions, the conditional probability function (CPF) was performed for each source using the source contributions estimated from the Unmix coupled with the surface wind direction data. It estimates the probability that a given source contribution from a certain wind direction will exceed a predetermined threshold criterion and is defined as:

$$CPF_{\Delta\theta} = m_{\Delta\theta} / n_{\Delta\theta},$$

where $m_{\Delta\theta}$ is the number of occurrence from wind sector $\Delta\theta$ that exceeded the threshold criterion and $n_{\Delta\theta}$ is the total number of data from the same wind sector. The sources are likely to be located in the directions that have high conditional probability values. In this study, 16 sectors were used ($\Delta\theta = 22.5°$) and calm wind (< 1 m s^{-1}) periods were excluded from the analysis. The threshold criterion was set at the upper 25th percentile value of the source contributions for each source.

For pollutant sources that are unknown, hybrid receptor models that incorporate wind trajectories, Potential Source Contribution Function (PSCF) [56] and Concentration Weighted Trajectory (CWT) [57], can be used to resolve source locations. The PSCF values can be interpreted as a conditional probability describing the spatial distribution of probable geographical source locations by analysis of trajectories arriving at the sampling site. Cells related to high values of the potential source contribution function are the potential source areas. However, the potential source contribution function maps do not provide an emission inventory of a pollutant but rather show those source areas whose emissions can be transported to the measurement site. Since the PSCF method is known to have difficulties in distinguishing strong from moderate sources, the CWT model that determines the relative significance of potential sources has been additionally performed. To estimate the likely source locations for regional transporting aerosols, the PSCF and CWT were calculated using the daily source contributions to PM_{10} concentration deduced from the Unmix and backward trajectories analysis.

Air mass back trajectories were computed by the HYSPLIT (HYbrid Single Particle Lagrangian Integrated Trajectory) model through interactive READY system [58,59]. Daily 48-hour back trajectories, started from Belgrade (44.804°, 20.478°) at 12:00 UTC each day, were evaluated for six different heights above the starting point at ground level (200, 350, 500, 750, 1000, and 1200 m). In addition, a study of airflow characteristics was performed using cluster analysis of 48-hour backward trajectories of air masses arriving above Belgrade. Since we were concerned on the directions of the trajectories, the angle distance between back trajectories has been used as the cluster model [60].

3.2. Health Risk Assessment

Elemental characterization of PM_{10} allowed for estimation of health risk associated with their trace element content. For that purpose the US EPA health risk assessment model [61,62] was used. In order to estimate the health risk, the information about toxicity of elements of interest for inhalation exposure route and an estimated daily exposure dose are required. Exposure is described by chronic daily intake, calculated as follows:

$$CDI = \frac{C \times IR \times EF \times ED}{BW \times AT}.$$

Here C is concentration of the chemical in the air, IR is inhalation rate, EF is exposure frequency, ED is exposure duration, BW is body weight, and AT is period over which the exposure is averaged. Values of these parameters used for assessment of health risk in this study are given in the Table 1.

In conducting health risk assessment, the carcinogenic and non-carcinogenic effects are considered separately. Parameter used as quantitative measure of toxicity is inhalation slope factor (*SF*) for carcinogenic elements, and inhalation reference dose (*RfD*) for non-carcinogenic elements. Their values are based on epidemiological and animal studies, and are obtained from US EPA Integrated Risk Information System (IRIS) [63] and Risk Assessment Information System (RAIS) [64] chemical toxicity databases.

For assessment of health risk due to exposure to carcinogenic elements the incremental lifetime cancer risk (*ILCR*) value is calculated, according to:

$$ILCR = CDI \times SF.$$

It represents an incremental probability of an individual developing a cancer over lifetime as a result of exposure to these elements through inhalation.

Table 1. Values of parameters used to calculate the chronic daily intake (CDI) of PM_{10}-bound trace elements

Parameter	Value
C	Chemical specific (mg m^{-3})
IR	20 $m^3 day^{-1}$
EF	350 days $year^{-1}$
ED	30 years
BW	70 kg
AT	70 x 365 days = 25 550 days for carcinogenic elements 30 x 365 days = 10 950 days for non-carcinogenic elements

Non-carcinogenic health risk due to exposure to analyzed elements is assessed by calculation of hazard quotient (*HQ*):

$$HQ = CDI/RfD.$$

In cases when inhalation *RfD* values were not available from IRIS and RAIS chemical toxicity databases, *HQ* was obtained by dividing concentration of the chemical with the minimum risk level (*MRL*) available from Agency for Toxic Substances and Disease Registry (ATSDR) database [65].

3.3. Results and Discussion

Results related to PM_{10} include total of 277 24-hour samples from two urban locations in Belgrade covering the period from 2003 to 2006. The analysis of mean annual PM_{10} mass concentrations revealed the increasing trend of concentrations from 63- μg m^{-3} (2003) to 89 μg m^{-3} (2006). The average concentration of 70μg m^{-3} for the whole period, as well as average concentration for each year, exceed the annual PM_{10} European Union air quality standard [66]. The highest mean trace elements concentration in PM_{10} was found for Al (3896 ng m^{-3}), followed by Fe (1408 ng m^{-3}) and Zn (1221 ng m^{-3}). Among other elements, the estimated mean levels of Cu, Pb, V, Mn, Ni, Cr and Cd were 85, 48, 38, 24, 22, 10 and 1.9 ng m^{-3} respectively. Average concentrations of Pb, Cd, V, and Mn in PM_{10} did not exceed values indicated by legislative directives while the average Ni concentration (22 ng m^{-3}) exceeded the related limit value of 20 ng m^{-3} according to the EC Directive [67].

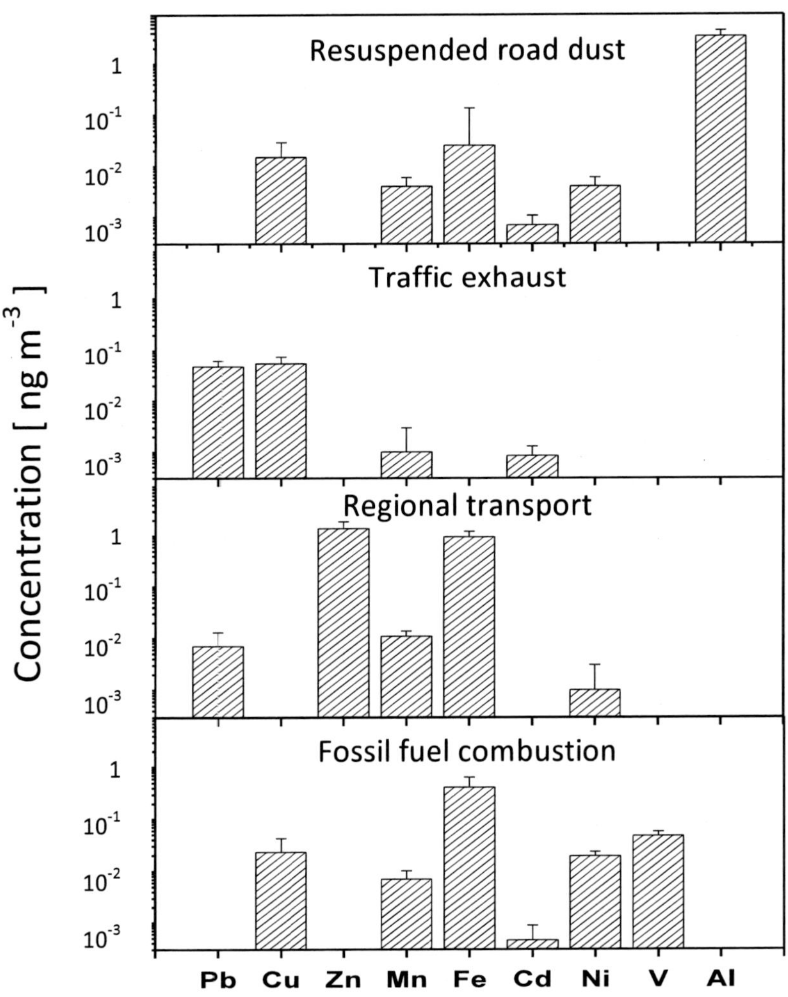

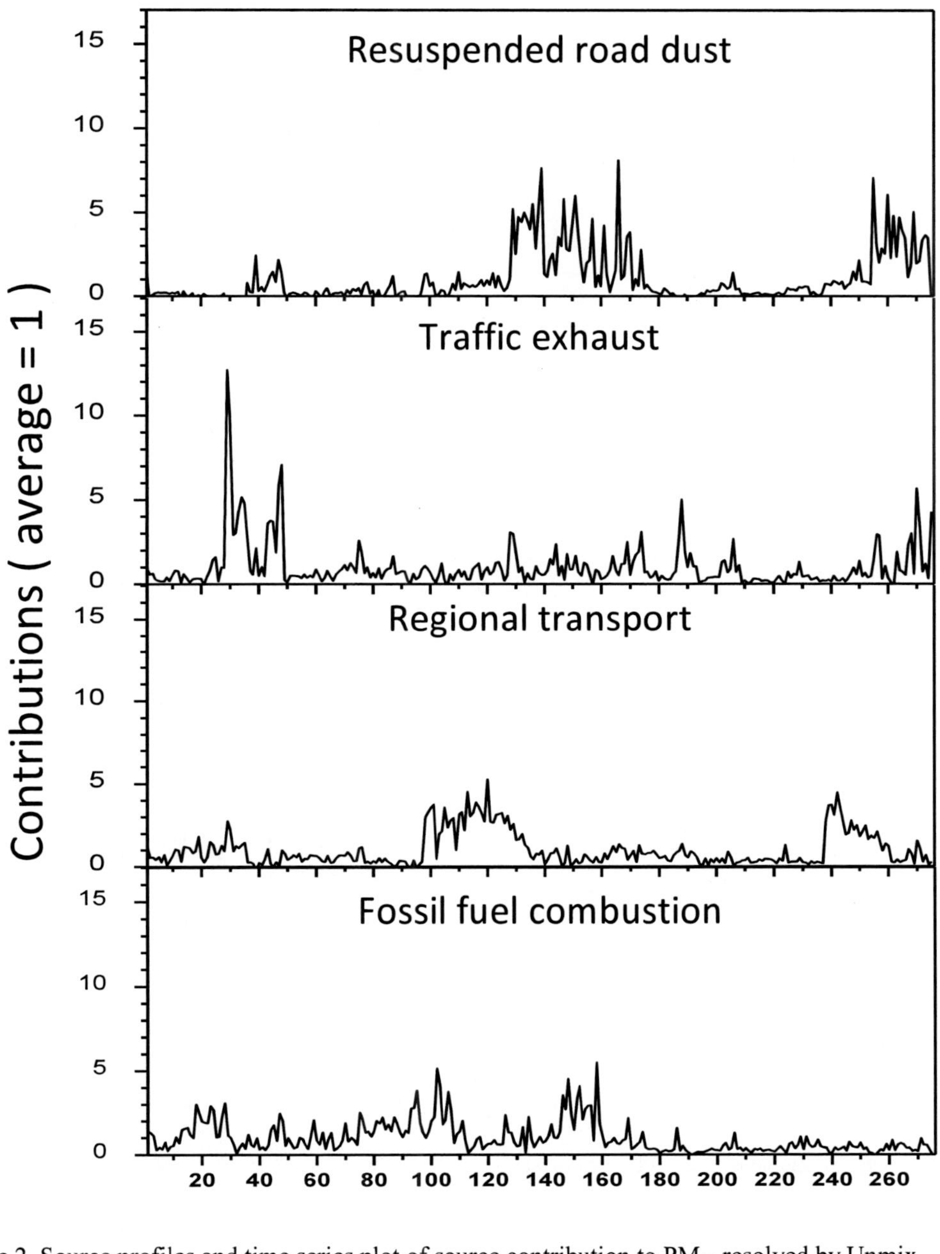

Figure 2. Source profiles and time series plot of source contribution to PM_{10} resolved by Unmix.

3.3.1. Source Apportionment

Unmix resolved four sources related to fossil fuel combustion (mainly consisting of V and Ni) with average contribution to PM_{10} concentration of 34%, regional transport mainly from steel and petrochemical industry (high loadings of Zn, Fe, and Mn) with average contribution of 26%, resuspended road dust (consisting of Al, Fe, Cu, Mn, Cd) with contribution of 19%, and traffic exhaust (high loadings of Pb and Cu) with contribution of 21% [68]. Source profiles and time series plot of the source contributions to PM_{10} concentration are presented in Figure 2.

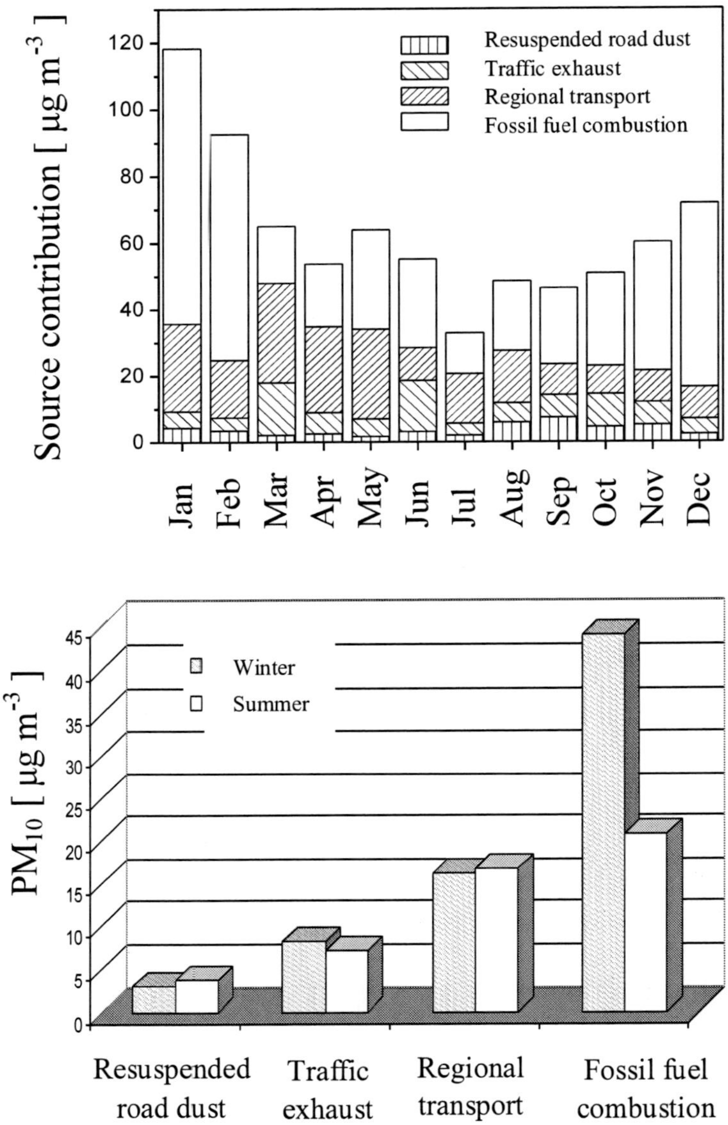

Figure 3. Monthly and seasonal average source contributions to PM_{10} resolved by Unmix.

The right panel shows the time series of contributions associated with the factor, where the contributions average to 1. Thus, a value of 2 in this plot means that the contribution for that factor to the receptor on that day (observation) is 2 times the average amount.

The seasonal variability of the identified sources was also examined and the results indicated strong dependence for fossil fuel combustion source. Figure 3 shows average source contribution, normalized to PM_{10} concentration, for each month and for winter and summer seasons. Opposite to resuspended road dust, it can be seen that traffic exhaust had slightly higher contribution during winter. Regional transport had no significant seasonal variability of contribution.

The CPF plots help identifying the impact of local point emissions to the Unmix-resolved sources and Figure 4 presents CPF plots for the studed results.

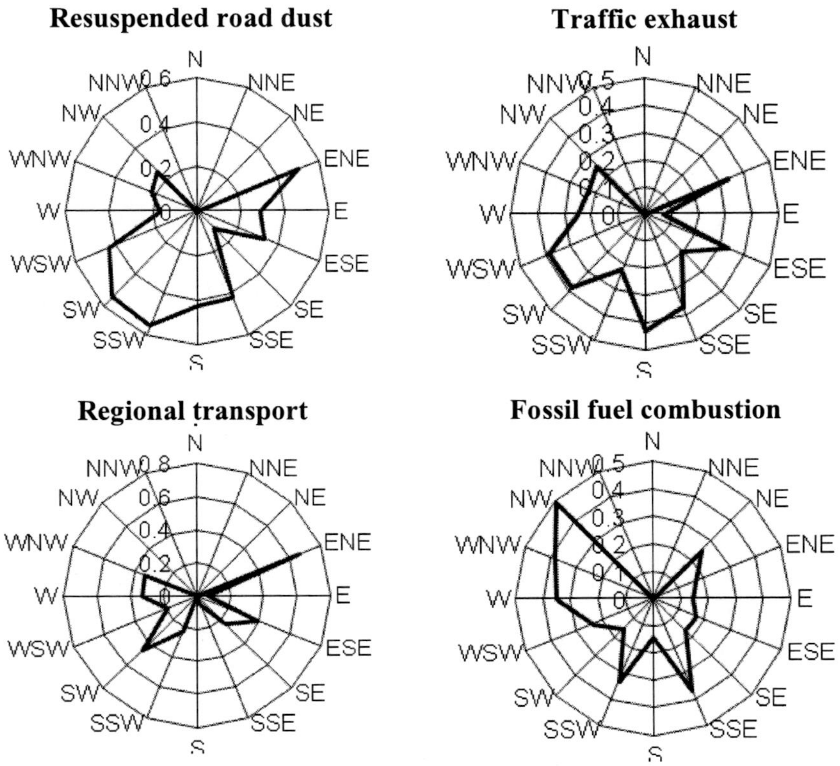

Figure 4. CPF plot for the average source contributions deduced from Unmix analysis.

It can be seen that the high contribution was coming from south-west except for fossil fuel combustion coming from north-west and moderate influence from south-east and north-east for all sources. In these directions there are several possible emission sources: coal-fired thermal power plants – "Nikola Tesla A and B", Obrenovac (west-southwest); "Kostolac A and B", Kostolac (east-southeast); "Kolubara", Veliki Crljeni (south-southwest); as well as the steel industry complex in Smederevo (east-southeast). In general, the directions of regional source location and possible transport pathway may not be in agreement with the local surface wind data used for the CPF plots. Therefore, hybrid receptor models have been used for further investigation of possible contribution of regional-transport. The Figures 5 and 6 show the results of PSCF and CWT analysis associated to Unmix-resolved transport and resuspended road dust. The results indicated that the strong potential probabilities are located in the south-west and east regions.

In addition, higher PSCF values are observed from north-east and north-west for resuspended dust. Similar results have been obtained using CWT model, indicated strong source emissions in west and east regions.

Additional insights into the nature of the PM_{10} sources are provided through a trajectory based evaluation and cluster analysis. A cluster number can be deduced through visual inspection and comparison of the mean-trajectory maps. We chose six clusters and their representative trajectories are shown on Figure 7, as well as the average PM_{10} concentration associated with each cluster.

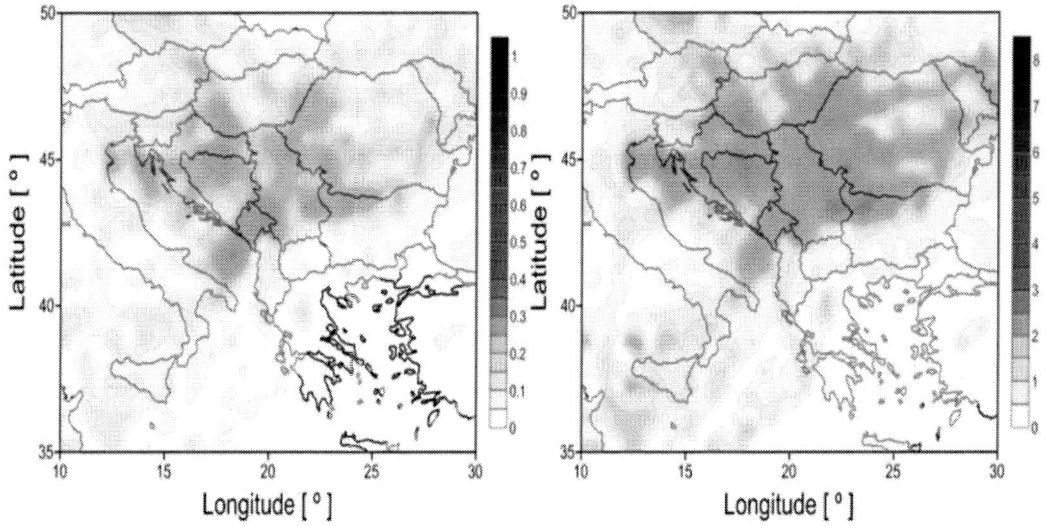

Figure 5. PSCF (left) and CWT (right, μg m^{-3}) map for the Unmix-resolved resuspended road dust.

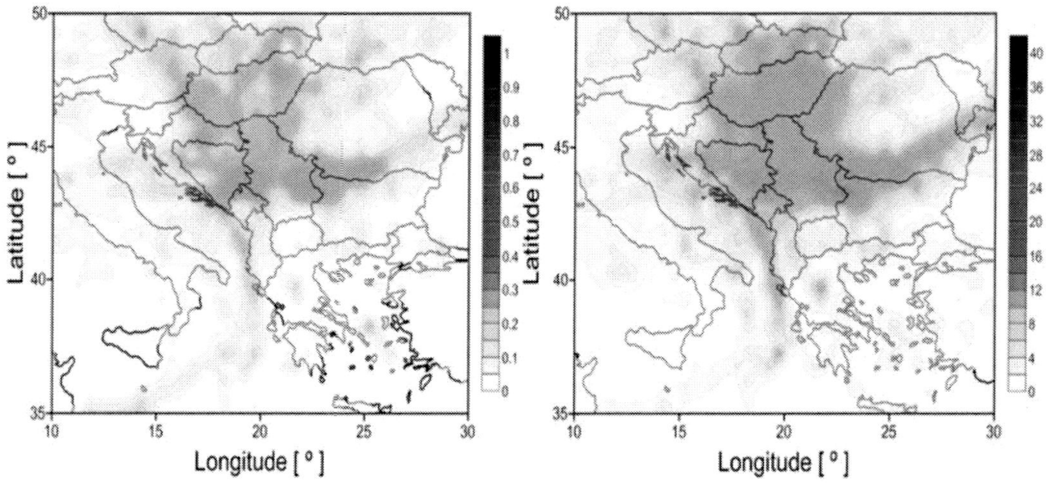

Figure 6. PSCF (left) and CWT (right, μg m^{-3}) map for the Unmix resolved regional transport.

Based on the analysis of the whole trajectory data set, the most frequently arriving directions are west-northwest (22%) and west-southwest (21%), followed by north-northeast (17%), north-northwest (16%), east (14%) and south (10%).

The results suggest that the highest PM_{10} concentrations were related to the west-southwest (63 μg m^{-3}) and south (58 μg m^{-3}) pathways. Along such pathways the high emissions could be located, which was indicated by high PSCF and CWT values.

3.3.2. Health Risk Results

The values of *ILCR* obtained for Pb, Cd, Ni, and Cr are in the range between 5×10^{-5} and 2×10^{-7}. The *ILCR* value of 10^{-6} corresponds to the exposure to unpolluted air. However, values lower than 10^{-4} are considered acceptable, and do not require risk reduction measures. The total ILCR of these four elements is 5.4×10^{-5}, significantly lower than the occurrence of cancer of about 4×10^{-3} in Belgrade, according to information from Belgrade Public Health Institute [69]. Among the considered elements the major contributors to cancer risk in Belgrade are Cr and Ni. However, it should be noted that total Cr concentration is used in *ILCR* estimate, due to lack of information on concentration of Cr(VI), which is carcinogenic. Note also that concentration of Ni exceeds the limit defined by EC directive, as mentioned above.

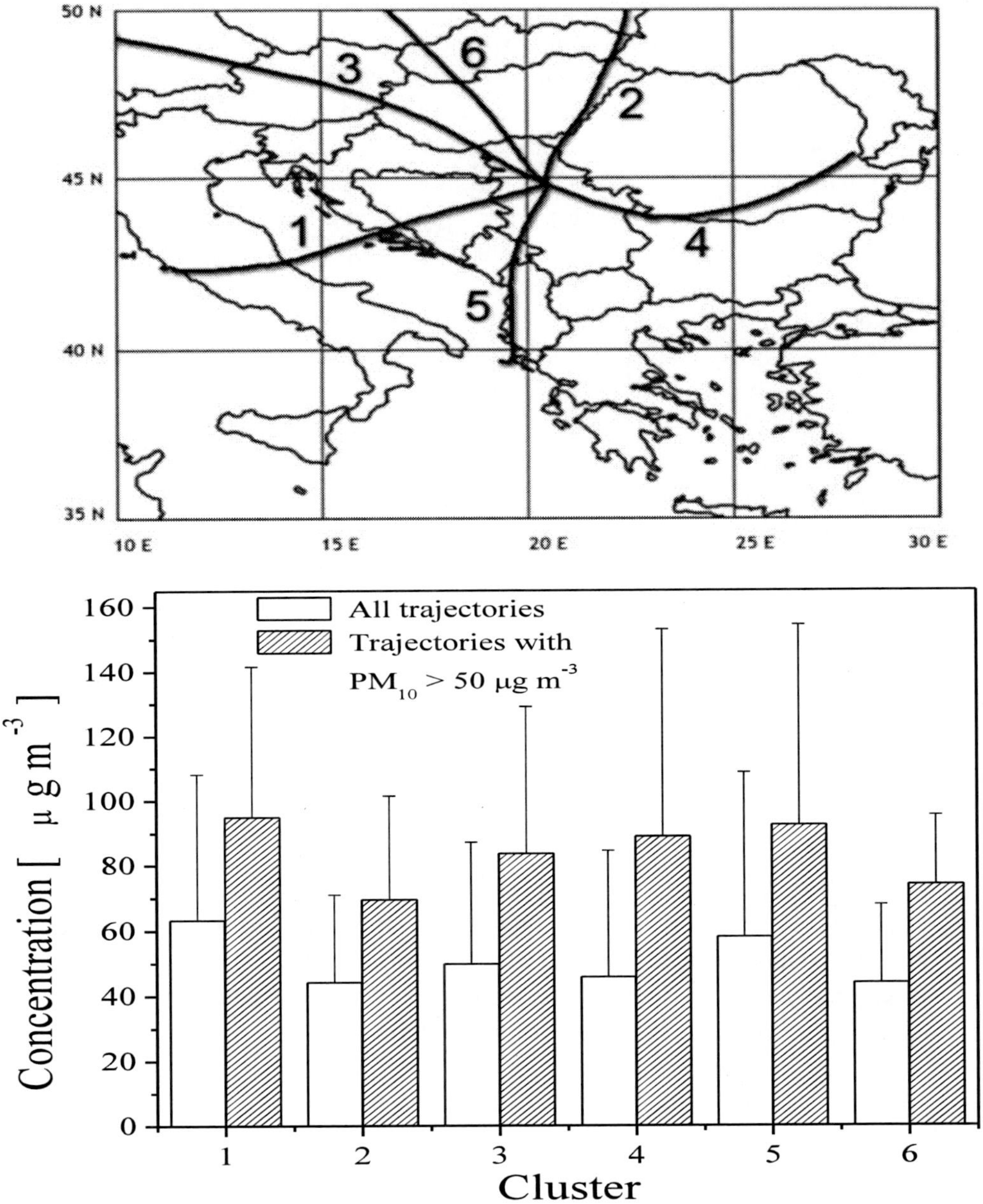

Figure 7. Mean trajectories representing grouping of 48-hour backward trajectories of air masses over Belgrade into the six classes and corresponding PM_{10} mass concentrations and standard deviation within classes.

All the estimated HQ values for analyzed PM_{10}-bound trace elements are lower than the nominal value 1. However, the hazard index (HI) obtained as the sum of HQ values for all analyzed elements exceeds the value of 1 (HI=1.6), indicating possible adverse non-carcinogenic health effects as a result of exposure to them through inhalation.

4. Active Biomonitoring of Trace Elements with Moss Bags

The first active moss biomonitoring experiment in Belgrade was carried out in the summer 2005. Since moss species from *Sphagnum* gender have been considered to be especially suitable for the monitoring of trace element air pollution, *Sphagnum girgensohnii* Russow was chosen for biomonitoring purposes in the Belgrade urban area. The moss bags technique initially involves the collection of moss species from its natural, relatively unpolluted habitat, than cleaning from extraneous matter, and finally suspending in nylon net in the field. Moss bags can be spherical and flat - horizontal or vertical [30]. The flat and vertical shape of bags was chosen with stretched thin layer of moss because it allows uniform collection of particles from all directions.

Moss bags are usually hanged freely by means of plastic string on aluminum holder placed at least 2 m above ground. However, moss bags tend to dry out and thus their efficiency in retaining elements varies depending on the environmental conditions, especially on humidity [70]. In order to prevent drying of mosses and maximize element collection efficiency, the exposed bags could be moistened with distilled water. In each of our experiments moss bags were exposed in parallel with and without irrigation, with the aim of testing the role of water in moss accumulation capacity.

Due to finite cation exchange capacity of moss and its differential affinities and retention capacities for cations, the exposure period should not be too long, preventing the saturation of all ion exchange sites. Also, in polluted environment, prolonged exposure times may exceed the ability of moss to retain insoluble airborne particles. Thus, in most surveys, exposure times of moss bags vary in range from few weeks to several months.

Several experiments were performed in Belgrade urban area in order to:

i. study the capacity of *Sphagnum girgensohnii* moss bags to accumulate trace element pollutants in an urban area under specific climatic conditions;
ii. examine the role of different humidity conditions (dry and irrigated - wet moss bags) on the accumulation efficiency of different elements, and
iii. investigate the relations between moss (dry and wet) trace element accumulation and exposure time looking for optimal exposure period.

4.1. Moss Sampling, Bag Preparation and Exposure

In each our experiment the same procedure of moss bag preparing for exposure was applied. An appropriate amount of *S. girgensohnii* was collected immediately before setting the experiments, in May 2005 and May 2007, from a pristine wetland area located near Dubna, Russia (φ = 56°44' N, λ = 37°09' E, H_s = 120 m). In laboratory, the moss was rigorously sorted to remove extraneous materials (soil particles and other foreign matter) from healthy green shots of the moss. Moss bags were prepared by packing about 3 g of air-dried moss in 10x10 cm in nylon net (1-mm mesh size). Different treatments, with and without irrigation, were applied in parallel to the exposed moss bags at all sampling sites (Figure 8).

Irrigation of moss bags was achieved by placing them on the top of a cellulose (100%) sponge with the bottom immersed in distilled water. The whole setup was placed in a polyethylene box (130 mm x 110 mm x 80 mm). Distilled water was added to the boxes at the intervals of several days depending on the meteorological conditions (precipitation and temperature).

The survey was carried out at three representative locations in heavy traffic areas of Belgrade (SP, BG, KP), from July 2005 to October 2006. Using specially constructed holders placed on platforms 5–10 m above the street level, the bags were exposed for five 3-month and two 6-month consecutive periods. Trace element accumulation was investigated in both, dry and wet moss bags [71]. At the same time, dry and wet atmospheric deposition was collected on a monthly basis using bulk collectors [72].

Figure 8. Experimental setup at IP suburban study site: A) dry, and B) wet moss bags.

The next study was carried out in a suburb of Belgrade (IP) as semi-urban location. The moss bags were exposed to the atmospheric deposition for different exposure periods (0.5 to 5 months) from June to November 2007 (Figure 8). Thus, the moss bags were exposed for ten consecutive 15-day periods [73].

4.2. Moss Sample Analysis

In laboratory, after the periods of exposure, the moss was removed from the nylon net, manually homogenized and dried in oven at 40˚C.

In the first experiment concentrations of 29 elements (Na, Al, Cl, K, Ca, Sc, Ti, V, Cr, Mn, Fe, Co, Ni, Zn, As, Se, Br, Rb, Sr, Sb, Cs, Ba, La, Ce, Sm, Tb, W, Th, and U) in the moss were determined by instrumental neutron activation analysis (INAA) with epithermal neutrons [74]. This method is nondestructive and samples were prepared in form of pellets (≈ 0.4 g) for further analysis. Short-term irradiation (2 min) was used for short-lived radionuclides (Al, V, and Mn). The long irradiation of 100 h was used to determine elements associated with long-lived radionuclides (Na, Cr, Fe, Ni, Zn, and As). Gamma-ray spectra were measured four times using a high-purity Ge detector after decay periods of 5 and 10 min following the short irradiation and after 3 and 20 days following the long irradiation, respectively. The low temperature during irradiation of samples (60 – 70°C) allows determination of elements present in the samples in volatile form. The quality control was performed by comparing the certified and measured values of element concentrations for three reference materials: Lichen (IAEA–336), Tomato Leaves (SRM–1573a) and Coal Fly Ash (SRM–1633b). Additionaly, the concentrations of Cu, Cd, and Pb in moss were analyzed by flame atomic absorption spectrometry (FAAS). Prior to this analysis about 0.5 g of moss sample (dry weight) was digested for 24 h in solution of 1 ml $HClO_4$ + 10 ml HNO_3 + 2 ml HCl on heating plate. The digested samples were diluted with distiled water to a total volume of 25 ml. Quality control was performed using standard reference material lichen 336 (IAEA).

In the second study moss samples were digested prior to further chemical analysis. The portions of approximately 0.5 g of moss (dry weight) were digested for 2 h in an UltraCLAVE microwave digester with 8 ml of concentrated nitric acid (Merck, Ultrapure) using a standard temperature program. After cooling to room temperature, the digested samples were diluted with demineralised water to a total volume of 60 ml. The concentrations of 49 elements (Be, Na, Mg, Al, Si, P, Cl, K, Ca, Sc, Ti, V, Cr, Mn, Fe, Co, Zn, Ga, As, Se, Rb, Sr, Y, Nb, Mo, Cd, Te, Cs, Ba, La, Ce, Pr, Sm, Eu, Tb, Dy, Er, Tm, Yb, Lu, Hf, Ta, W, Hg, Tl, Pb, Bi, Th, and U) were determined by inductively-coupled plasma mass spectrometry (ICP-MS). Quality control was performed using two international moss reference samples [75] and tea leaves GBW-07605 (Institute of Geophysical and Geochemical Exploration, Langfang, China).

Basic statistics was calculated to investigate differences and relationships between element concentrations measured in the unexposed moss, dry and wet moss bags and between different periods of exposure.

4.3. Results and Discussion

4.3.1. Trace Element Content in Moss Bags versus Exposure Time

According to some research, the higher the element concentrations in the moss before the exposure are, the lower the relative accumulation of the elements is [76]. Thus, initial trace element concentrations in the unexposed moss material were determined, to be considered for the experiments described in this chapter. The amount of trace elements found in *S. girgensohnii* before exposure was compared with data previously reported for baseline concentrations within the same moss genus from relatively unpolluted areas. For the majority of the elements, the initial values measured in the *S. girgensohnii* in this study were significantly lower than those reported for the *Sphagnum sp.* moss originating from some other background sites [34,77,78], indicating that this pristine sampling site in Dubna (Russia) is a very suitable background (unpolluted) area. Generally, the initial element concentrations in the unexposed moss obtained by both analytical methods (INAA and ICP-MS) in our studies [71,73,79] were at the same level having in mind that moss was collected from the pristine area at different times (May 2005 and May 2007).

The influence of the initial values in the moss may be expressed in terms of a relative accumulation factor (RAF):

$$RAF = (C_{exposed} - C_{initial})/C_{initial},$$

where $C_{exposed}$ is the content of the element in the moss samples after exposure, while $C_{initial}$ is the content of the same element determined before the exposure. In our study, high RAF values in dry moss bags were obtained for some elements after the 6-month exposure periods (Table 2).

The highest values of RAF were obtained for elements such as V, Sb, Cr, As, Pb, Cu, Zn, etc. due to their very low initial concentrations in the moss and possible high ambient air concentration level during moss bags exposure.

Table 2. Relative accumulation factors (RAF) of elements in dry moss samples after the 6 months of exposure

RAF	Element
>40	V, Sb
20 – 40	Cr, As
10 – 20	Al, Sc, Cu, Pb, La, Sm, Th
5 – 10	Ca, Fe, Zn, Cd, Ce, U
1 – 5	Co, Ni, Se, Br, Sr, Ba
0 – 1	Na, Mn
< 0	Cl, K, Rb, Cs

In general, V is indicator of oil combustion, as well as Ni. Antimony, Cr, Al, Fe, Zn, La, and Th probably represent road dust resuspension, which includes soil dust mixed with traffic related particles. Lead, Cr, Zn, Ba, and Br may be associated with vehicle exhaust emission, and As with tram trails. Calcium comes mostly from weathered buildings material [1]. Results of some trace element instrumental monitoring studies indicate that resuspended road dust and fossil fuel combustion are the most significant sources of the measured elements in air of the urban Belgrade area [46, 49]. Particulate matter is a predominate form of trace elements emissions in urban areas and uptake capacity of mosses mainly depends on passive physico-chemical entrapment and adsorption of elements on cell walls. Studies on the capture of atmospheric particles by moss have demonstrated that standardized active biomonitoring with moss bags provided better capture efficiency of particles over 20 μm diameter (sedimentable particles) less influenced by abiotic conditions like wind speed. Therefore, it was suggested that particles trapped by bryophytes may be a major source of poorly water-soluble elements and that moss content can reflect recent environmental conditions for dry and coarse depositions, especially for active biomonitoring experiments in highly polluted areas [80].

However, relocation of mosses from the natural habitat represents a strong stress for them, causing damage to cell membranes and consequently leakage of some elements. Thus, loss of some, mostly physiologically active, elements from the moss tissue was evident in our experiments after different periods of exposure [73,79]. Sodium, P, Cl, K, Rb, Cs, and Ta concentrations in exposed moss (dry) were below the initial concentrations, i.e. for these elements negative RAF values were obtained (Table 2).

Loss of some of the above-mentioned elements from the moss tissue was also observed in other studies employing moss bags [76,77,81,82]. Leakage of K and other physiologically active elements such as Na, Cl and P, may reflect cell damage [81,82] or displacement of some ions by other incoming ions with greater affinities for the cation exchange sites on moss cell walls [82]. Culicov *et al.* [76] also found that some elements (Cl, K, Rb, Cs) were depleted in bags of *S. girgensohnii* and ascribed this to a combination of two factors: the accumulation of heavy metals and desiccation/hydration cycles. Additionally, results from our studies imply that the loss of these elements (P, K, Na, Rb, Cs, and Cl) is less pronounced for wet moss bags than for dry ones due to the maintenance of moss humidity during the exposure [73].

In the case of Mn, Mg, and Br there was no significant difference in concentration before and after the moss exposure. Positive but still low RAF values were observed for these elements (Table 2; Table 3). Moreover, the data corresponding to the 6-month exposure indicate that these elements were released from the moss at increased rates. Consequently, the moss bag technique appears to be of limited value for these elements as well.

In the second experiment (set up in suburban area of Belgrade), for majority of determined 49 elements, the increase of concentrations in the *S. girgensohnii* moss were observed even for 15-day period of bags exposure (Figure 9). However, significant element concentrations in comparison with the initial element content were measured for the 1-month moss exposure periods.

Also, constant and linear increase of the concentrations during exposure was observed for most elements showing significant difference in accumulation of elements between one, two and three months of exposure. The same trend was observed for four and five months of exposure (Table 3) [73]. All this suggests that *S. girgensohnii* moss species is sufficiently sensitive trace element biomonitor.

Table 3. Average concentrations (μg g^{-1} d.w.) of the 49 elements assayed in the initial moss, dry and wet moss bags of *S. girgensohnii* exposed from June 2007 for 1-5 months

Element		Initial moss		Months of exposure									
				1		2		3		4		5	
		N=5	SD	N=30	SD	N=28	SD	N=30	SD	N=3	SD	N=3	SD
Be	dry mb	**0.011**	0.002	**0.017**	0.004	**0.025**	0.005	**0.032**	0.004	**0.036**	0.003	**0.059**	0.013
	wet mb	**0.011**	0.002	**0.022**	0.006	**0.028**	0.003	**0.034**	0.005	**0.040**	0.004	**0.049**	0.006
Na	dry mb	**112**	8.4	**92**	17.8	**98**	17.2	**81**	20.3	**65**	1.1	**49**	3.5
	wet mb	**112**	8.4	**158**	55.2	**134**	51.3	**132**	55.8	**96**	9.5	**73**	5.6
Mg	dry mb	**1,905**	315	**1,734**	207	**1,944**	228	**2,009**	210	**2,120**	103	**1,769**	68
	wet mb	**1,905**	315	**2,156**	242	**2,219**	211	**2,287**	210	**1,997**	218	**1,752**	56
Al	dry mb	**167**	18	**368**	64	**540**	106	**716**	100	**836**	5	**1,321**	234
	wet mb	**167**	18	**465**	110	**588**	75	**761**	77	**955**	33	**1,153**	21
Si	dry mb	**340**	93	**693**	234	**981**	169	**1,276**	175	**1,489**	45	**2,088**	193
	wet mb	**340**	93	**827**	192	**1,061**	137	**1,381**	175	**1,785**	165	**2,022**	98
P	dry mb	**1,793**	92	**947**	208	**854**	130	**714**	73	**712**	20	**719**	31
	wet mb	**1,793**	92	**1,171**	311	**1,155**	203	**1,109**	192	**1,109**	213	**1,200**	85
Cl	dry mb	**541**	53	**139**	140	**102**	34	**63**	31	**86**	34	**42**	11
	wet mb	**541**	53	**442**	260	**322**	165	**342**	171	**259**	87	**236**	64
K	dry mb	**9,574**	443	**4,312**	1,761	**2,930**	1,141	**1,386**	658	**713**	29	**669**	33
	wet mb	**9,574**	443	**5,133**	1,768	**3,463**	1,613	**2,722**	1,255	**2,203**	716	**1,996**	281
Ca	dry mb	**3,638**	430	**4,467**	661	**5,395**	768	**6,297**	662	**7,027**	648	**8,264**	1,366
	wet mb	**3,638**	430	**5,842**	808	**6,718**	582	**7,748**	839	**7,650**	768	**8,040**	520
Sc	dry mb	**0.02**	0.002	**0.06**	0.01	**0.09**	0.02	**0.13**	0.02	**0.15**	0.00	**0.26**	0.05
	wet mb	**0.02**	0.002	**0.07**	0.02	**0.10**	0.02	**0.14**	0.02	**0.18**	0.01	**0.21**	0.01
Ti	dry mb	**4.25**	0.26	**7.52**	1.52	**10.95**	2.32	**14.72**	2.28	**16.44**	0.54	**27.66**	4.99
	wet mb	**4.25**	0.26	**7.21**	1.60	**8.93**	1.35	**11.92**	1.69	**16.84**	1.19	**20.01**	1.11
V	dry mb	**0.52**	0.01	**1.13**	0.43	**1.78**	0.78	**2.42**	0.69	**2.72**	0.21	**5.91**	0.87
	wet mb	**0.52**	0.01	**1.16**	0.32	**1.58**	0.28	**2.08**	0.32	**2.56**	0.25	**3.44**	0.07
Cr	dry mb	**0.26**	0.02	**0.83**	0.39	**1.53**	0.73	**2.16**	0.73	**2.33**	0.13	**5.32**	1.44
	wet mb	**0.26**	0.02	**1.08**	0.28	**1.78**	0.51	**2.61**	0.46	**3.35**	0.33	**4.36**	0.10
Mn	dry mb	**341.4**	32.0	**379.3**	72.9	**423.9**	63.5	**416.1**	49.7	**451.0**	52.7	**363.1**	39.4
	wet mb	**341.4**	32.0	**413.0**	60.3	**391.3**	50.2	**360.1**	57.4	**290.7**	21.9	**259.2**	44.5
Fe	dry mb	**150**	20	**349**	69	**500**	118	**665**	113	**763**	13	**1,229**	264
	wet mb	**150**	20	**441**	110	**535**	63	**693**	84	**812**	20	**1,067**	43
Co	dry mb	**0.17**	0.03	**0.34**	0.06	**0.44**	0.11	**0.53**	0.08	**0.55**	0.03	**0.76**	0.16
	wet mb	**0.17**	0.03	**0.44**	0.10	**0.50**	0.08	**0.55**	0.08	**0.59**	0.03	**0.65**	0.08
Zn	dry mb	**23.7**	0.6	**29.4**	3.3	**36.0**	5.5	**44.7**	6.3	**47.8**	0.1	**62.5**	7.2
	wet mb	**23.7**	0.6	**35.3**	4.6	**41.6**	5.3	**49.8**	6.3	**63.2**	14.2	**69.4**	7.3
Ga	dry mb	**0.041**	0.003	**0.089**	0.021	**0.135**	0.032	**0.188**	0.031	**0.224**	0.006	**0.371**	0.071
	wet mb	**0.041**	0.003	**0.107**	0.025	**0.143**	0.020	**0.193**	0.027	**0.245**	0.019	**0.302**	0.017
As	dry mb	**0.10**	0.004	**0.19**	0.07	**0.25**	0.10	**0.31**	0.09	**0.40**	0.04	**0.79**	0.10
	wet mb	**0.10**	0.004	**0.33**	0.12	**0.53**	0.25	**0.78**	0.40	**0.95**	0.14	**1.79**	0.14
Se	dry mb	**0.099**	0.008	**0.113**	0.007	**0.137**	0.016	**0.145**	0.012	**0.167**	0.012	**0.219**	0.022
	wet mb	**0.099**	0.008	**0.134**	0.023	**0.151**	0.019	**0.186**	0.028	**0.215**	0.014	**0.277**	0.007
Rb	dry mb	**50.5**	3.7	**25.3**	8.4	**18.1**	6.8	**8.5**	4.5	**3.6**	0.2	**2.8**	0.2
	wet mb	**50.5**	3.7	**29.3**	9.2	**20.2**	9.2	**15.7**	6.4	**12.1**	3.9	**12.0**	1.5
Sr	dry mb	**6.9**	1.0	**10.3**	1.7	**12.4**	2.3	**14.6**	1.8	**15.1**	1.4	**18.0**	3.5
	wet mb	**6.9**	1.0	**15.2**	2.8	**18.3**	2.0	**21.1**	2.1	**19.6**	1.3	**20.0**	3.3
Y	dry mb	**0.053**	0.007	**0.128**	0.021	**0.184**	0.038	**0.242**	0.034	**0.277**	0.010	**0.465**	0.113
	wet mb	**0.053**	0.007	**0.151**	0.034	**0.191**	0.030	**0.261**	0.038	**0.349**	0.017	**0.424**	0.009
Nb	dry mb	**0.017**	0.002	**0.026**	0.004	**0.034**	0.006	**0.041**	0.005	**0.038**	0.001	**0.053**	0.006
	wet mb	**0.017**	0.002	**0.018**	0.008	**0.015**	0.008	**0.015**	0.013	**0.029**	0.006	**0.038**	0.005
Mo	dry mb	**0.126**	0.008	**0.117**	0.015	**0.149**	0.024	**0.175**	0.020	**0.210**	0.021	**0.342**	0.011
	wet mb	**0.126**	0.008	**0.135**	0.026	**0.171**	0.109	**0.186**	0.023	**0.225**	0.031	**0.259**	0.010
Cd	dry mb	**0.179**	0.019	**0.276**	0.048	**0.307**	0.042	**0.342**	0.028	**0.361**	0.042	**0.430**	0.077
	wet mb	**0.179**	0.019	**0.296**	0.028	**0.329**	0.104	**0.312**	0.029	**0.334**	0.079	**0.320**	0.015
Te	dry mb	**0.0010**	0.0004	**0.0016**	0.0003	**0.0020**	0.0003	**0.0024**	0.0004	**0.0024**	0.0002	**0.0040**	0.0003
	wet mb	**0.0010**	0.0004	**0.0025**	0.0014	**0.0028**	0.0006	**0.0034**	0.0007	**0.0030**	0.0004	**0.0041**	0.0008

Table 3. (Continued)

Element		Initial moss		Months of exposure									
				1		2		3		4		5	
		N=5	SD	N=30	SD	N=28	SD	N=30	SD	N=3	SD	N=3	SD
Cs	dry mb	**0.205**	0.009	**0.137**	0.027	**0.128**	0.018	**0.119**	0.013	**0.119**	0.004	**0.160**	0.028
	wet mb	**0.205**	0.009	**0.151**	0.032	**0.131**	0.025	**0.139**	0.019	**0.150**	0.014	**0.169**	0.007
Ba	dry mb	**19.5**	3.9	**30.6**	6.2	**35.9**	7.7	**39.9**	6.7	**44.1**	7.2	**48.0**	14.3
	wet mb	**19.5**	3.9	**34.6**	5.7	**36.4**	4.8	**37.1**	5.0	**33.1**	4.7	**34.9**	8.6
La	dry mb	**0.099**	0.011	**0.251**	0.044	**0.362**	0.075	**0.483**	0.070	**0.564**	0.015	**0.997**	0.309
	wet mb	**0.099**	0.011	**0.301**	0.068	**0.381**	0.064	**0.537**	0.166	**0.695**	0.043	**0.817**	0.022
Ce	dry mb	**0.177**	0.016	**0.461**	0.080	**0.687**	0.143	**0.918**	0.127	**1.060**	0.047	**1.925**	0.592
	wet mb	**0.177**	0.016	**0.566**	0.136	**0.720**	0.122	**1.040**	0.331	**1.343**	0.121	**1.581**	0.050
Pr	dry mb	**0.019**	0.001	**0.048**	0.009	**0.073**	0.015	**0.099**	0.014	**0.116**	0.005	**0.206**	0.061
	wet mb	**0.019**	0.001	**0.060**	0.014	**0.078**	0.014	**0.113**	0.037	**0.147**	0.012	**0.175**	0.008
Sm	dry mb	**0.013**	0.001	**0.033**	0.006	**0.051**	0.012	**0.069**	0.011	**0.083**	0.003	**0.144**	0.033
	wet mb	**0.013**	0.001	**0.042**	0.010	**0.054**	0.008	**0.079**	0.022	**0.107**	0.012	**0.127**	0.006
Eu	dry mb	**0.005**	0.001	**0.01**	0.002	**0.02**	0.003	**0.02**	0.004	**0.02**	0.002	**0.04**	0.009
	wet mb	**0.005**	0.001	**0.01**	0.003	**0.02**	0.002	**0.02**	0.005	**0.03**	0.002	**0.04**	0.003
Tb	dry mb	**0.0018**	0.0002	**0.0045**	0.0008	**0.0068**	0.0013	**0.0091**	0.0013	**0.0106**	0.0004	**0.0183**	0.0045
	wet mb	**0.0018**	0.0002	**0.0056**	0.0014	**0.0072**	0.0012	**0.0103**	0.0022	**0.0139**	0.0013	**0.0165**	0.0007
Dy	dry mb	**0.0057**	0.0004	**0.0138**	0.0024	**0.0203**	0.0041	**0.0271**	0.0039	**0.0317**	0.0002	**0.0536**	0.0121
	wet mb	**0.0057**	0.0004	**0.0166**	0.0040	**0.0213**	0.0033	**0.0298**	0.0050	**0.0408**	0.0024	**0.0489**	0.0015
Er	dry mb	**0.005**	0.0005	**0.012**	0.002	**0.018**	0.004	**0.024**	0.004	**0.027**	0.000	**0.046**	0.011
	wet mb	**0.005**	0.0005	**0.015**	0.003	**0.019**	0.003	**0.026**	0.003	**0.034**	0.002	**0.041**	0.001
Tm	dry mb	**0.0007**	0.0001	**0.0016**	0.0003	**0.0023**	0.0005	**0.0030**	0.0004	**0.0035**	0.0000	**0.0060**	0.0013
	wet mb	**0.0007**	0.0001	**0.0018**	0.0004	**0.0024**	0.0004	**0.0032**	0.0004	**0.0044**	0.0003	**0.0052**	0.0003
Yb	dry mb	**0.0045**	0.0006	**0.0096**	0.0017	**0.0141**	0.0030	**0.0191**	0.0027	**0.0215**	0.0004	**0.0371**	0.0074
	wet mb	**0.0045**	0.0006	**0.0113**	0.0024	**0.0147**	0.0023	**0.0201**	0.0024	**0.0268**	0.0012	**0.0318**	0.0017
Lu	dry mb	**0.0006**	0.00004	**0.0013**	0.0002	**0.0020**	0.0004	**0.0027**	0.0004	**0.0031**	0.0000	**0.0052**	0.0011
	wet mb	**0.0006**	0.00004	**0.0016**	0.0003	**0.0021**	0.0003	**0.0029**	0.0003	**0.0037**	0.0002	**0.0045**	0.0003
Hf	dry mb	**0.0030**	0.0004	**0.0054**	0.0008	**0.0077**	0.0017	**0.0101**	0.0015	**0.0108**	0.0005	**0.0171**	0.0032
	wet mb	**0.0030**	0.0004	**0.0051**	0.0019	**0.0041**	0.0025	**0.0047**	0.0036	**0.0106**	0.0048	**0.0132**	0.0043
Ta	dry mb	**0.0009**	0.00001	**0.0009**	0.0001	**0.0009**	0.0001	**0.0008**	0.0001	**0.0005**	0.0001	**0.0007**	0.0001
	wet mb	**0.0009**	0.00001	**0.0004**	0.0003	**0.0002**	0.0002	**0.0003**	0.0001	**0.0007**	0.0006	**0.0005**	0.0001
W	dry mb	**0.027**	0.002	**0.042**	0.010	**0.061**	0.016	**0.071**	0.014	**0.083**	0.010	**0.118**	0.010
	wet mb	**0.027**	0.002	**0.038**	0.010	**0.044**	0.024	**0.039**	0.025	**0.083**	0.010	**0.078**	0.011
Hg	dry mb	**0.027**	0.001	**0.036**	0.005	**0.044**	0.006	**0.045**	0.004	**0.045**	0.002	**0.051**	0.006
	wet mb	**0.027**	0.001	**0.030**	0.004	**0.030**	0.006	**0.030**	0.005	**0.027**	0.003	**0.033**	0.004
Tl	dry mb	**0.014**	0.002	**0.029**	0.005	**0.031**	0.007	**0.031**	0.005	**0.026**	0.003	**0.025**	0.002
	wet mb	**0.014**	0.002	**0.030**	0.007	**0.029**	0.008	**0.027**	0.006	**0.026**	0.008	**0.034**	0.005
Pb	dry mb	**1.69**	0.17	**4.62**	0.98	**5.65**	1.36	**7.22**	1.27	**8.59**	0.56	**11.71**	2.86
	wet mb	**1.69**	0.17	**5.34**	0.99	**6.27**	1.01	**7.56**	1.31	**8.95**	0.82	**10.90**	0.70
Bi	dry mb	**0.008**	0.0001	**0.015**	0.007	**0.022**	0.008	**0.029**	0.007	**0.031**	0.002	**0.055**	0.009
	wet mb	**0.008**	0.0001	**0.016**	0.004	**0.019**	0.003	**0.025**	0.006	**0.031**	0.003	**0.036**	0.003
Th	dry mb	**0.016**	0.0004	**0.041**	0.008	**0.063**	0.015	**0.087**	0.013	**0.104**	0.008	**0.188**	0.040
	wet mb	**0.016**	0.0004	**0.050**	0.015	**0.070**	0.019	**0.099**	0.016	**0.147**	0.019	**0.165**	0.011
U	dry mb	**0.007**	0.001	**0.015**	0.003	**0.023**	0.007	**0.031**	0.006	**0.036**	0.001	**0.062**	0.013
	wet mb	**0.007**	0.001	**0.019**	0.004	**0.024**	0.003	**0.033**	0.003	**0.043**	0.003	**0.050**	0.003

N=number of separate samples (moss bags); SD - standard deviation; Initial moss (N=5); 1 month (N=30); 2 months (N=28); 3 months (N=30); 4 months (N=3); 5 months (N=3).

Only a few quantitative comparisons of biomonitoring methods with the physical or chemical measurements of atmospheric deposition have been published so far [83,84,85]. Moreover, the exact relationship between the element content in moss and actual atmospheric deposition has not yet been well understood, even though some studies have given evidence of possible quantitative conversion with unsedimentable dry deposited particles (< 0.8 μm) [33].

In the first moss bags survey in Belgrade, the 3-month and 6-month exposure periods covered two main seasons: "non-heating" (May–October) and "heating" (November–April) corresponding to the official heating season. Significant seasonal variation in dry and wet moss bags was observed for some of the elements, especially for V and Ni.

The concentrations (RAF values) of these elements were several times higher in winter than in summer. Such results were not unexpected because V and Ni are markers for oil and coal combustion [1]. Seasonal variation of these two elements in Belgrade urban area was also confirmed by analysis of bulk atmospheric deposition samples [72] as well as in the above-mentioned PM_{10} samples [86]. High correlations between the element measured in bulk deposits and the same element concentration in moss bags were found for some of eleven elements that were compared: V ($r = 0.97$), Cu ($r = 0.83$), Ni ($r = 0.79$), As ($r = 0.73$), and Al ($r = 0.71$). Slightly lower correlations for Pb, Fe, Cr, and Zn, were obtained and there were no correlations for Cd and Mn [71]. Poor correlations for some elements in moss samples and bulk deposits indicate more complex mechanisms of element accumulation in mosses, due to their morphological and physiological characteristics. Total element content of moss may be considered as a result of the balance between an input from wet and dry deposition, and output determined by: *(i)* washing of particulate materials by rain; *(ii)* leaching of some ions due to precipitation (especially acid rain); *(iii)* cation displacement, depending on their relative affinities for binding sites and concentrations; *(iv)* cellular damage due to environmental stress [23]. Thus, it is not accurate to regard moss as a mere passive filter or collector.

4.3.2. Trace Element Concentrations in Dry and Wet Moss Bags

Due to a change of environment, the exposure in bags presumably causes physiological stress in the moss. The element accumulation by moss bags depends on various environmental conditions related to the climate and to the period of the year when the exposure occurred [87]. Growth and nutrient uptake may be strongly suppressed by intermittent desiccation [36]. Permanently irrigated moss bags buffer the effect of external meteorological conditions and can keep a physiological activity during a longer exposure [80]. The wet moss bags (living) may at least partly incorporate the elements in their tissues by active accumulation, being less susceptible to modification by precipitation and thus better reflecting the atmospheric pollution conditions [88]. Still, after a rain event low physiological activity of dry (unwatered) moss is recovered rapidly to a normal level [36, 80]. Many elements are added partly with precipitation in dissolved form, and the continued exposure to precipitation may cause some of the initially particle-bound matter to dissolve and become available to active and passive processes on the moss surface affecting dissolved species. Thus, Adamo *et al.* [77] compared the accumulation of metals in moss during dry and wet seasons, and stated that the uptake was affected by meteorological conditions for Al, Co, Cr, Cu, Fe, Mg, Mo, Pb, Ti, and Zn. The results from our studies, performed in Belgrade urban area, clearly indicate that continuous irrigation of the exposed moss could improve the accumulating capacity of some elements.

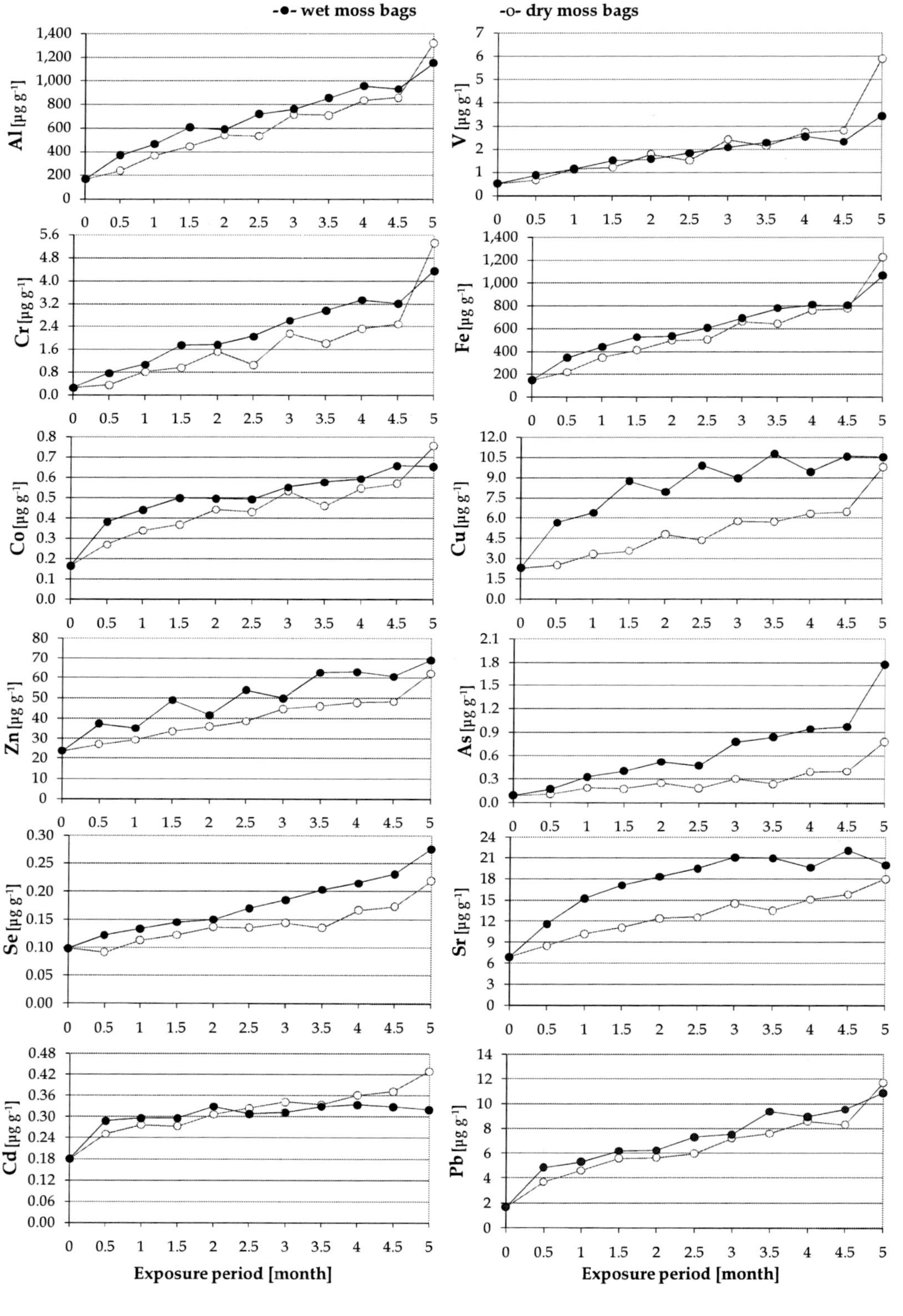

Figure 9. The concentration (µg g^{-1}) of Al, V, Cr, Fe, Co, Cu, Zn, As, Se, Sr, Cd, and Pb in wet and dry moss bags from June to October 2007 in steps of 15 days.

In the first study, the higher relative uptake of the irrigated moss bags was evident for Al, Cr, Fe, Zn and Sr during all five consecutive 3-month exposure periods (which include whole year) and the wet/dry ratio of moss bags was always over 1 [79]. In the second study, comparison of element accumulation by dry and wet moss bags with increasing the exposure periods (1-2-3 months) indicated higher cumulative properties of wet moss for many of the studied elements: Sr, Se, As, Zn, Ca, Mg, Te ($p < 0.001$), and Al, Cr, Sm, Si, Be, Ce, Co, Mo, Dy, La ($p < 0.05$). However, several elements such as Hf, Nb, Ti, and W were more accumulated in dry moss bags (Table 3; Figure 9). For some elements, such as Ba, Bi, Eu, and Ga, accumulation levels were approximately of the same order in wet and dry moss bags. Also, the previously mentioned loss of the elements P, K, Na, Rb, Cs, and Cl from moss tissue during exposure was lower ($p < 0.001$) in wet moss [73].

The *Sphagnum* species are capable of absorbing large amount of water and keeping it within its water-carrying large cells [88]. Therefore, permanent irrigation of moss enables binding of heavy metal ions for ion exchange sites to a higher extent than in dry moss. Also, water-soluble elements from deposited particles became more available for uptake by moss.

Figure 9 presents the concentrations of accumulated elements (Al, V, Cr, Fe, Co, Zn, As, Se, Sr, Cd, Hg, and Pb) by dry and wet moss bags with time in steps of 15 days, as well as their initial values. These elements are typical tracers for urban air pollution (motor vehicles, oil burning, coal combustion, waste incineration, etc.). An increasing trend of accumulation with time is evident for all of these elements for both types of moss bags. The elements were accumulated linearly over periods of four months. During the fifth month, the concentrations increased rapidly in both types of moss bags, reflecting the start of heating season (coal combustion tracers: Cd, Fe, Hg, Pb, Se, V, Zn) in Belgrade and additional emission sources in the winter [46]. In the last exposure period (October), higher enrichment was evident for most of the above elements in dry moss bags than in wet. It may be related to passive phenomena of surface adsorption of fly ash as a dominant residue generated by coal combustion. Coarse and sparingly soluble particles were more efficiently trapped by dry moss bags.

CONCLUSION

For last ten years in the urban area of Belgrade instrumental monitoring and active moss biomonioring of airborne trace elements have been conducted in the consecutive experiments. At the several sites in the heavy traffic area the analysis of PM_{10} mass concentrations, as well as trace element concentrations in PM_{10} were performed. The highest mean trace elements concentration in PM_{10} was found for Al, followed by Fe, and Zn. Average concentrations of Pb, Cd, V, and Mn in PM_{10} did not exceed values indicated by legislative directives, while the average Ni concentration exceeded related limit value according to the EC Directive.

The Unmix receptor model resolved four major emission sources: fossil fuel combustion, regional transport, resuspended road dust, and traffic exhaust. The impact from the local point sources was explored by coupling wind direction data with the daily source contributions deduced from the Unmix. The effects of regional sources were also found using PSCF and CWT analysis for the Unmix-resolved source contributions. Based on the cluster analysis of the 48-hour backward trajectories six classes of backward trajectories were generated. The highest PM_{10} mass concentrations were found along the west-southwest and south pathways.

Along such pathways, the high emissions could probably be located, which were indicated by high PSCF and CWT values. This study suggested that the regional transport should be considered to control the level of trace elements content in PM_{10}.

The results of assessment of health risk due to exposure to PM_{10}-bound trace elements suggest potential adverse non-carcinogenic health effects, and possible need for risk management measures. The obtained ILCR values suggest the level of carcinogenic health risk in urban area of Belgrade that is considered acceptable.

In addition to instrumental measurements of trace elements in the urban air, indirect monitoring method, i.e. active moss biomonitoring of the trace elements, were performed. The selected moss species, *Sphagnum girgensohnii*, was packed in bags (the half of all exposed bags were watered), and exposed to atmospheric deposition at selected urban sites. Both, dry and wet, moss bags showed constant and linear increasing trend of concentrations for a majority of the studied elements: Al, Si, Ca, Ti, V, Cr, Fe, Co, Zn, Ga, As, Se, Sr, Cd, Te, Ba, Hg, Pb, and several rare-earth elements during 1-5 months of exposure. However, higher accumulation in wet moss bags was evident for the following elements: Sr, Se, As, Zn, Ca, Mg, Te, Al, Cr, Sm, Si, Be, Ce, Co, Mo, Dy, La, Hf, Nb, Ti, and W. Some, physiologically active elements (P, K, Na, Rb, Cs, and Cl) were depleted from moss tissue during exposure and thus this moss species could not be used for ambient monitoring of these elements.

Very similar trends of element accumulation for dry and wet moss bags indicates that both methods of exposure may be used as appropriate tools for biomonitoring of airborne trace elements. In case of heavily polluted urban areas, 1-month moss bags exposure period could be recommended as sufficient. However, in case of relatively low air pollution wet moss bags could be a better choice by showing higher accumulation capacity, presumably due to prolonged vitality of the moss. Exposure period of up to 5 months may be preferable for moss bags biomonitoring of trace elements in case of very low ambient concentration. The results of our survey generally confirmed that the use of *S. girgensohnii* moss bags is a simple, sensitive and inexpensive way of obtaining extensive information on deposition levels of atmospheric trace elements in urban areas.

In future research on air pollution it is very much of interest to apply interdisciplinary approach, integrating instrumental monitoring and biomonitoring of emitted elements, and predicting distribution of trace elements through modeling.

ACKNOWLEDGMENTS

This paper was realized as a part of the project No III43007 financed by the Ministry of Education and Science of the Republic of Serbia within the framework of integrated and interdisciplinary research for the period 2011-2014. The authors gratefully acknowledge the NOAA Air Resources Laboratory (ARL) for the provision of the HYSPLIT transport and dispersion model and READY website (http://www.arl.noaa.gov/ready.php) used in this publication.

REFERENCES

[1] Pacyna, J.M.; Pacyna, E.G. 2001. *Environ. Rev.* 9, 269-298.

[2] Seinfeld, J.H.; Pandis, S.N. *Atmospheric Chemistry and Physics from Air Pollution to Climate Change,* John Wiley and Sons, New York, 1998.

[3] Vukmirović, Z.; Tošić, I.; Tasić, M.; Rajšić, S. 2004. *Meteorol. Atmos. Phy.* 85 (4), 265-273.

[4] Viana, M.; Kuhlbusch, T.A.J.; Querola, X.; Alastueya, A.; Harrison R.M.,; Hopke, P.K..; Winiwartere W.; Vallius, M.; Szidat, S.; Prévôt, A.S.H.; Hueglin C.; Bloemenk, H.; Wåhlin, P.; Vecchi, R.; Miranda A.I.; Kasper-Giebl, A.; Maenhaut W.; Hitzenberger R. 2008. *Aeros. Sci.* 39, 827–849.

[5] Bargagli, R. *Trace Elements in terrestrial Plants: An Ecophysiological Approach to Biomonitoring and Biorecovery*, Springer-Verlag Berlin, Heidelberg, N. Y., 1998.

[6] Kabata-Pendias, A., Mukherjee, A. *Trace Elements from Soil To Human,* Springer, 2007.

[7] Nriagy, J.O.; Pacyna, J.M. 1988. *Nat.* 333, 134-139.

[8] Alcamo, J. In: *Coping with Crisis in Eastern Europe's Environment*, Ed. Alcamo, J. IIASA Book, Parthenon Pub. Grp., Carnforth, 1992. 87- 101.

[9] Azimi, S.; Ludwig, A.; Thevenot, D.R.; Colin, J-L. 2003. *Sci. Tot. Environ.* 308, 247-254.

[10] Morselli, L.; Olivieri, P.; Brusori, B.; Passarini, F. 2003. *Environ. Pollut.* 124, 457-469.

[11] Pope C. A.; Dockery D. W.; Schwartz J. 1995. *Inhal. Toxic.* 7, 1-18.

[12] Wolterbeek, B. 2002. *Environ. Pollut.* 120, 11–21.

[13] Szczepaniak, K.; Biziuk, M. 2003. *Review. Environ. Res.* 93, 221-230.

[14] Tomašević, M.N.; Vukmirović, Z.B.; Rajšić, S.; Tasić, M.; Stevanović, B. 2005. *Chemosphere* 61, 753 – 760.

[15] Tomašević, M.; Vukmirović, Z.; Rajšić, S.; Tasić, M.; Stevanović, B. 2008. *Environ. Monit. Assess.* 137, 393–401.

[16] Aničić, M.; Spasić, T.; Tomašević; M.; Rajšić, S.; Tasić, M. 2011. *Ecol. Ind.* 11, 824–830.

[17] Šucur, K.; Aničić, M.; Tomašević, M.; Antanasijević, D.; Perić-Grujić, A.; Ristić, M. 2010. *J. Serb. Chem. Soc.* 75 (10), 1453-1461.

[18] Tomašević, M.; Aničić, M.; Jovanović, Lj.; Perić-Grujić, A.; Ristić, M. 2011. *Ecol. Ind.* 11, 1689–1695.

[19] Gorelova, S.V.; Frontasyeva, M.V.; Yurukova, L.; Coskun, M.; Pantelica, A.; Saitanis, C.J.; Tomašević, M.N.; Aničić, M.P. 2011. *Agrochimica* 55 (2), 65-84.

[20] Rühling, Å.; Tyler, G. 1968. *Bot. Not.* 121, 321-342.

[21] Harmens, H.; Norris, A.D.; Koerber, R.G.; Buse, A.; Steinnes, E.; Ruhling, A. 2008. *Environ. Pollut.* 151, 368-376.

[22] Schröder, W.; Pesch R.; Englert, C.; Harmens, H.; Suchara, I.; Zechmeister, H.G.; H.G.,Thöni, L.; Maňkovská, B.; Jeran ,Z.; Grodzinska, K.; Alber, R. 2008. *Environ. Pollut.* 151, 377-388.

[23] Brown, H.D. In: *Bryophyte ecology*, Eds. A. J. E. Smith, London: Chapman and Hall, 1982, 383–444.

[24] Buck, G.W.; Brown, D.H. 1979. *Ann. Bot.* 44, 265–277.

[25] Honegger, R. 1995. *Can. J. Bot.* 73, 569–587.

[26] Tyler, G. 1990. *Bot. J. Linnen. Soc.* 104, 231-253.

[27] Fernandez, A.J.; Aboal, R.J.; Carballeira, A. 2004. *Ecotoxicol. Environ. Saf.* 59, 76-83.

[28] Aničić, M.; Frontasyeva, M.V.; Tomašević, M.; Popović, A. 2006. *Environ. Monit. Assess.* 129, 207-219.

[29] Goodman, T.G.; Roberts, M.T. 1971. *Nat.* 231, 287-292.

[30] Little, P.; Martin, H.M. 1974. *Environ. Pollut.* 6, 1–19.

[31] Martin, M.H.; Coughthrey, P.J. Biological monitoring of heavy metal pollution. *Land and Air.* Applied Science Publishers, 1982, 311–336.

[32] Hynninen,V. 1986. *Ann. Bot. Fennici* 23, 83–90.

[33] Vasconcelos, M.T.S.D.; Tavares, H.M.F. 1998. *Sci. Tot. Environ.* 212, 11-20.

[34] Culicov, O.A.; Yurukova, L. 2006. *J. Atmos. Chem.* 55, 1–12.

[35] Richardson, D.H.S., 1981. *The biology of mosses.* London: Blackwell Scientic Publications.

[36] Bates, W.; Shaw, A.J. *Mineral nutrition and substratum ecology in Bryophyte Biology* Edited by Jonathan and Goffinet Bernard, Cambridge (University Press), 2000, pp. 248.

[37] Zechmeister, H.G.; Grodzińska, K.; Szarek-Łukaszewska, G. In: *Bioindicators and biomonitors*, B.A. Markert, A.M. Breure, H.G. Zechmeister, Elsevier Science, 2003.

[38] Clymo, S.R. 1963. *Ann. Bot.* 27, 309-324.

[39] Makinen, A. 1977. *Suo.* 28, 79–88.

[40] Ratcliffe, M.J. 1975. *Atmos. Environ.* 9, 623-629.

[41] Goodman, T.G.; Smith, S.; Parry, R.D.G.; Inskip, J.M. 1974. *Natl. Soc. Clean Air, Annu. Conf.* 41, 1-11.

[42] Samara, C.; Kouimtzis, Th.; Tsitouridou, R.; Kanias, G.; Simeonov, V. 2003. *Atmos. Environ.* 37, 41-54.

[43] Mijić, Z.; Tasić M.; Rajšić S.; Novaković, V. 2009. *Atmos. Res.* 92, 420-426.

[44] Tasić, M.; Đurić-Stanojević, B.; Rajšić, S.F.; Mijić, Z.; Novaković, V.T. 2006. *Acta Chim. Slov.* 53, 401-405.

[45] Rajšić, F.S.; Tasić, D.M.; Novaković, T.V.; Tomašević, N.M. 2004. *Environ. Sci. Pollut. Res.* 11, 158-164.

[46] Rajšić, S.; Mijić, Z.; Tasić, M.; Radenković, M.; Joksić, J. 2008. *Environ. Chem. Lett.* 6, 95–100.

[47] Tasić, D.M.; Rajšić, F.S.; Novaković, V.T.; Mijić, R.Z.; Tomašević, M.N. 2005. *Phys. Scripta.* T118, 29-30.

[48] Hopke, K.P. 2003. *J.Chemometr.* 17, 255-265.

[49] Mijić, Z.; Rajšić, S.; Žekić, A.; Perišić, M.; Stojić, A.; Tasić, M. In: *Air quality*, Ed. Ashok Kumar, Sciyo Publisher, Croatia, 2010, 143-167.

[50] Marković, M.D.; Marković, A.D.; Jovanović, A.; Lazić, L.; Mijić, Z. 2008. *Environ. Monit. Assess.* 145, 349–359.

[51] U.S. Environmental Protection Agency, 2007. EPA Unmix Version 6.0, available from *http://www.epa.gov/heasd/products/unmix/unmix.html.*

[52] Henry, R.C. 1997. *Chemometr. Intell. Lab.* 37, 525–530.

[53] Henry, R.C. 2003. *Chemometr. Intell. Lab.* 65, 179 – 189.

[54] Poirot, R.L.; Wishinski, P.R; Hopke, P.K; Polissar, A.V. 2001. *Environ. Sci. Technol.* 35, 4622-4636.

[55] Song, Y.; Dai, W.; Shao, M.; Liu, Y.; Lu, S.; Kuster, W.; Goldan, P. 2008. *Environ. Pollut.* 156, 174-183.
[56] Ashbaugh, L.L.; Malm, W.C.; Sadeh, W.Z. 1985. *Atmos. Environ.* 19, 1263-1270.
[57] Hsu, Y-K.; Holsen, M.T.; Hopke, K.P. 2003. *Atmos. Environ.* 37, 545-562.
[58] Draxler, R.R.; Rolph, G.D. 2010. *HYSPLIT (HYbrid Single-Particle Lagrangian Integrated Trajectory)* Model access via NOAA ARL READY Website (http://ready.arl.noaa.gov/HYSPLIT.php)
[59] Rolph. G.D. 2010. *Real-time Environmental Applications and Display sYstem (READY) Website (http://ready.arl.noaa.gov),* NOAA Air Resources Laboratory, Silver Spring, MD.
[60] Wang, Y.Q.; Zhang, X.Y.; Draxler, R. 2008. *Environ. Modell. Softw.* 24, 938-939.
[61] US EPA. United States Environmental Protection Agency. 1989. EPA/540/1-89/002. NTIS PB90-155581
[62] US EPA. United States Environmental Protection Agency. 1992. EPA/600/Z-92/001.
[63] IRIS. (2012) Integrated Risk Information System. http://www.epa.gov/ iris/
[64] RAIS. (2009) The Risk Assessment Information System. *http://rais.ornl. gov/tools/tox_profiles.html.*
[65] ATSDR. (2010). Agency for Toxic Substances and Didease Registry, Minimal Risk Levels (MRLs). *www.atsdr.cdc.gov/mrls.*
[66] EC (2008) Directive 2008/50/EC.
[67] EC (2004) Directive 2004/107/EC.
[68] Mijić, Z.; Tasić, M.; Rajšić, S.; Stojić, A. 2011. Book of Abstracts, *3rd International WeBIOPATR Workshop and Conference.*
[69] BPHI. (2009). Belgrade Public Health Institute. *http://www.zdravlje. org.rs/statistika/*
[70] Al-Radady, A.S.; Davies, B.E.; French, M.J. 1993. *Sci. Tot. Environ.* 133, 275-283.
[71] Aničić, M.; Tasić, M.; Frontasyeva, M.V.; Tomašević, M.; Rajšić, S.; Mijić, Z., Popović, A. 2009. *Environ. Pollut.* 157, 673–679.
[72] Mijić Z.; Stojić A.; Perišić M.; Rajšic S.; Tasić M.; Radenković M.; Joksić J. 2010. *Atmos. Environ.* 44, 3630-3637.
[73] Aničić, M.; Tasić, M.; Rajšić, S.; Popović, A.; Frontasyeva, M.V.; Lierhagen, S.; Steinnes, E. 2009. *J. Haz. Mat.* 171, 182-188.
[74] Frontasyeva, M.V.; Pavlov, S.S. 2000. *JINR Preprint*, E14- 2000- 177, Dubna.
[75] Steinnes, E.; Rühling, Å.; Lippo, H.; Mäkinen, A. 1997. *Accred. Qual. Assur.* 2, 243 – 249.
[76] Culicov, O.A.; Mocanu, R.; Frontasyeva, M.V.; Yurukova, L.; Steinnes, E. 2005. *Environ. Monit. Assess.* 108, 229-240.
[77] Adamo, P.; Giordano, S.; Vingiani, S.; Castaldo, C.R.; Violante, P. 2003. *Environ. Pollut.* 122, 91–103.
[78] Djingova, R.; Kuleff, I.; Markert, B. 2004. *Ecol. Res.* 19, 3–11.
[79] Aničić, M.; Tasić, M.; Frontasyeva, M.V.; Tomašević, M.; Rajšić, S.; Strelkova, L.P.; Popović, A.; Steinnes, E. 2009. *Environ. Chem. Lett.* 7, 55–60.
[80] Amblard-Gross, G.; Férard, J.F.; Carrot, F.; Bonnin-Mosbah, M.; Maul, S.; Ducruet, J.M.; Coddeville, P.; Béguinel, P.; Ayrault, S. 2002. *Environ. Pollut.* 120, 47–58.
[81] Brown, D.H.; Wells, J.M. 1990. *Ann. Bot.* 66, 641–647.
[82] Makholm, M.M.; Mladenoff, J.D. 2005. *Environ. Monit. Assess.* 104, 1-18.
[83] Sloof, E.J. 1995. *Atmos. Environ.* 29 (1), 11-20.

[84] Thöni, L.; Schnyder, N.; Krieg, F. 1996. *Fresenius J. Anal. Chem.* 354, 703-708.

[85] Berg, T.; Steinnes, E. 1997. *Environ. Pollut.* 98 (I), 61-71.

[86] Todorović, D.; Popović, D.; Rajšić, S.; Tasić, M. In: *Environmental Research Trends*, Editor: Marvin A. Cato, 271-301, 2007.

[87] Tretiach, M.; Adamo, P.; Bargagli, R.; Baruffo, L.; Carletti, L.; Crisafulli, P.; Giordano, S.; Modenesi, P.; Orlando, S.; Pittao, E. 2007. *Environ. Pollut.* 146, 380–391.

[88] Astel, A.; Astel, K.; Biziuk, M. 2008. *Environ. Sci. Pollut. Res.* 15, 41-50.

In: Trace Elements ISBN: 978-1-62081-376-8
Editors: Diego A. De Leon and Paloma Raquel Aragon

Chapter 2

FLOW OPTOSENSING APPLIED TO THE ANALYSIS OF TRACE ELEMENTS

Antonio Ruiz-Medina [*] ***and Eulogio J. Llorent-Martínez***
Department of Physical and Analytical Chemistry,
Faculty of Experimental Sciences, University of Jaén,
Campus Las Lagunillas, Jaén, Spain

ABSTRACT

The determination of trace amounts of inorganic ions is of growing interest in several fields including environmental analysis, process control, biology, medicine, etc., and there is now a particular need for simple and fast field analytical tests. In contrast to well-established laboratory methods, such as atomic absorption or emission spectrometry, mass spectrometry or voltammetry, chemical sensors have attracted a great deal of interest in the last decade because of their great potential for in the field, *in situ*, continuous and remote, if required, applications.

Flow-through optosensing emerged from the combination of Solid-Phase Spectroscopy (SPS) and Flow Injection Analysis (FIA). In SPS, an active solid support is used to pre-concentrate the target analyte and its spectroscopic detection is performed on the same solid support microbeads. The most remarkable features of SPS are its high sensitivity (it can be easily increased one hundred-fold due to the concentration of the target analyte on the solid beads) and selectivity (the interaction between the target chemical species and the solid phase beads is usually different from that one of co-existing species). In addition, the implementation of SPS in flow methods of analysis adds the inherent advantages of flow analysis, improving the analytical methods in terms of rapidity, commodity, automation, less consumption of reagents, etc. In these flow optosensors, the (micro) zone of the solid phase, where the signal is continuously monitored, is surrounded by a continuous stream flowing through it. The separation and retention of the species of interest on the solid phase takes place in the detection area itself and the spectroscopic determination is simultaneously performed.

In this chapter, we will first describe the fundamentals of SPS from its origin, including some relevant examples. After that, we will focus on the versatility and

[*] Tel: (+34)953-212759; Fax: (+34)953-212940; E-mail: anruiz@ujaen.es.

potential of flow-through optosensing for the analysis of trace elements in a wide variety of samples related to human health, such as clinical samples (drugs, urine, and serum), food or environmental waters, emphasizing the most recent advances and future trends in this field.

1. Introduction

Certain elements, although present in minute amounts in the tissues, are essential nutrients. They perform functions indispensable for the maintenance of life, growth or reproduction. Inadequate intakes may impair cellular and physiological function and often cause illness. They can be divided into those essential for higher animals, possibly essential and non-essential (toxic elements could be included here too). It must be taken into account that any essential trace element becomes toxic when it enters an organism in large amounts [1]. Therefore, the reliable quantification of these elements at trace and ultra-trace levels in a wide variety of samples is mandatory. In this sense, the characteristics required for an analytical method to be successful are, at least, high sensitivity and selectivity, robustness and high-reliability. Other characteristics such as automation, low-cost or simplicity are also desirable.

In the last few years, atomic spectroscopy techniques, such as flame- and graphite furnace atomic absorption spectroscopy, inductively coupled plasma-atomic emission spectroscopy or inductively coupled plasma–mass spectrometry have been widely used for the determination of trace ions in different samples, such as drugs, foods or waters [2-4]. Although they offer high selectivity and sensitivity, they are usually time-consuming and high-cost techniques. Special attention should be paid to the implementation of these sensors in automated systems, which will be the main topic of this chapter [5].

The inception of flow analysis has proved to be an excellent tool for handling solutions and, consequently, performing methods related to wet chemical analysis. Especially prominent among flow techniques is Flow Injection Analysis (FIA), proposed in 1975 by Ruzicka and Hansen [6], which initiated a field of research that would, over more than three decades result in close to thousand of publications in the international scientific literature. In general, in this technique, the sample is introduced by using a propelling system and processed under reproducible conditions, allowing its isolation from the environment and, therefore, avoiding external contamination [7]. Inherent features of FIA are low consumption of solutions, good accuracy and precision of results, high sample throughput, and overall robustness.

The main aims of recent developments in flow analysis have been to progressively achieve lower reagents consumption, higher repeatability, complete automation of the analytical procedure and miniaturization. In this sense, Sequential Injection Analysis (SIA) [8], Multicommutated Techniques [9] or Lab-On-Valve [10] are also widely employed due to their favorable characteristics. Although all of them offer inherent outstanding advantages, they have to be complemented by selective and sensitive detection techniques, additional separation devices and clean-up steps when analyzing complex matrices or determining more than one analyte. In this sense, Solid-Phase Spectroscopy (SPS) plays an important role. In SPS, the species of interest is sorbed on an appropriate solid support and its detection is performed directly on the solid phase; using this approach, great selectivity and sensitivity are achieved. These inherent features make SPS an interesting option for its implementation in automatic flow systems, as it could be used for the retention of the reagents, the target analyte or even the product of the reaction. The coupling between SPS and flow systems lead to the so called flow-through optosensors (FTO), which are versatile tools for the analysis of one or several analytes in complex samples. Only conventional FIA and Multicommutated Flow Injection Analysis (MCFIA) have been employed in the development of FTO for the analysis of trace elements. These systems present a high versatility and, therefore, are suitable for a wide range of applications, including *in situ* measurements. MCFIA is one of the main multicommutated methodologies that are available in scientific literature [7].

After explaining in detail the concepts of SPS and FTO, the most important innovations and applications in trace elements determination in different samples will be described.

2. Trace Elements and Micronutrients

Four elements (oxygen, carbon, hydrogen and nitrogen) account for 96% of living matter. In humans and other mammals, approximately 30 elements have known physiological activities, including the so called trace elements. To be considered essential, a trace element must be present in all healthy tissues of all organisms, its concentration must be relatively constant and withdrawal produces similar structural and physiological abnormalities in different species, which are prevented or reversed by addition of the element. The elements fulfilling these requirements are, at least, arsenic, chromium, cobalt, copper, fluoride, iodine, iron, manganese, molybdenum, nickel, selenium, silicon, tin, vanadium and zinc. Other potential candidates include barium, boron, bromide or strontium [1].

Trace elements originate in the Earth and cannot be made by living organisms. Plants obtain them from the soil, and most of the elements (macronutrients or trace elements) in our diets come directly from plants or indirectly from animal sources. They may also be present in the water we drink, but this varies with geographic locale. Trace elements from plant sources may also vary from place to place, because soil mineral content varies geographically. Trace elements in airborne dust arise from the Earth's surface and from the oceans, and also descend from sources beyond the atmosphere. The terrestrial aerosols are also raised by winds, by the eruption of volcanoes and by industrial activities. Large particles promptly return to the surface by sedimentations, but the disposal of smaller particles may be more widespread. The erosion of soils can increase the levels of some elements in natural waters. Heavy metals are used in industry and are consequently discharged into the environment. These elements occur in relatively low concentration in nature, but anthropogenic activities have contributed to increase their concentration in the environment.

Trace elements are needed by the human body in very small quantities (generally less than 100 mg/day), as opposed to those elements considered macronutrients (sodium, calcium, magnesium, potassium, chlorine, etc), which are required in larger quantities. These elements are essential components of biological structures, but at the same time they can be toxic at concentrations beyond those necessary for their biological functions. In addition, the toxicity can be extended to other non-essential elements of similar atomic characteristics that can mimic the reactivity of a trace element. To deal with this essentiality/toxicity duality, biological systems have developed the ability to recognize a metal, and deliver it to the target without allowing the metal to participate in toxic reactions [11].

Proteins are primarily responsible for such recognition and transport. However, a proper control of the levels of these elements in pharmaceuticals, drinking waters or food is mandatory to avoid potential health problems.

Estimation of dietary exposure is crucial for risk assessment, and possibly for determining the relationships between adverse effects observed in humans and dietary exposure to particular substances. Dietary exposure to elements such as cadmium, mercury, arsenic and lead has been associated with toxic and adverse health effects. In contrast, some trace elements are essential micronutrients that need to be consumed in adequate amounts to maintain normal physiological functions [12]. They may be toxic when taken in excessive amounts or may lead to deficiencies when taken in insufficient amounts. Therefore, control of these elements in the human body is necessary. The most important essential trace elements will be briefly discussed, showing the Recommended Dietary Allowance (RDA) for adults in Table 1.

Iron is a key element in the metabolism of almost all living organisms. In humans, iron is an essential component of hundreds of proteins and enzymes, such as hemoglobin and myoglobin, which are involved in the transport and storage of oxygen. Hemoglobin is the primary protein found in red blood cells and represents about two thirds of the body's iron. The vital role of hemoglobin in transporting oxygen from the lungs to the rest of the body is derived from its unique ability to acquire oxygen rapidly during the short time it spends in contact with the lungs and to release oxygen as needed during its circulation through the tissues. Myoglobin functions in the transport and short-term storage of oxygen in muscle cells, helping to match the supply of oxygen to the demand of working muscles. Its deficiency causes anemia, which results when there is inadequate iron to support normal red blood cell formation.

Zinc plays important roles in growth and development, the immune response, neurological function, and reproduction. On the cellular level, the function of zinc can be divided into three categories: catalytic, structural, and regulatory. 1) Nearly 100 different enzymes depend on zinc for their ability to catalyze vital chemical reactions. 2) Zinc plays an important role in the structure of proteins and cell membranes. 3) Zinc finger proteins have been found to regulate gene expression by acting as transcription factors (binding to DNA and influencing the transcription of specific genes).

Table 1. RDA of the main essential trace elements

Copper	900 μg
Iodine	150 μg
Iron	8-18 mg
Manganese	1.8-2.3 mg
Molibdenum	45 μg
Selenium	55 μg
Zinc	8-11 mg

Copper is an essential trace element for humans and animals. In the body, copper shifts between the cuprous (I) and cupric (II) forms, though the majority of the body's copper is in the II form. The ability of copper to easily accept and donate electrons explains its important role in oxidation-reduction (redox) reactions and in scavenging free radicals. Copper is a critical functional component of a number of essential enzymes known as cuproenzymes, which are involved in cellular energy production, in the formation of strong and flexible connective tissue or to catalyze normal function of the brain and nervous system.

Iodine, a non-metallic trace element, is required by humans for the synthesis of thyroid hormones and is therefore essential for normal thyroid function. Iodine deficiency is an important health problem throughout much of the world. Most of the earth's iodine is found in oceans, and iodine content in the soil varies with region. The older an exposed soil surface, the more likely the iodine has been leached away by erosion.

Humans and animals require selenium for the function of a number of selenium-dependent enzymes, also known as selenoproteins. During selenoprotein synthesis, selenocysteine is incorporated into a very specific location in the amino acid sequence in order to form a functional protein.

Molybdenum functions as a cofactor for a number of enzymes that catalyze important chemical transformations in the global carbon, nitrogen, and sulfur cycles.

Thus, molybdenum-dependent enzymes are not only required for human health, but also for the health of our ecosystem.

A number of manganese-activated enzymes play important roles in the metabolism of carbohydrates, amino acids, and cholesterol. Wound healing is a complex process that requires increased production of collagen. Manganese is required for the activation of prolidase, an enzyme that functions to provide the amino acid proline, for collagen formation in human skin cells.

3. Solid Phase Spectroscopy

In 1976 K. Yoshimura proposed the combined use of an active solid support to preconcentrate the analyte (or its reaction product with a chromogenic reagent) with the direct measurement of the light absorption of the species of interest sorbed on the solid phase, being this methodology called SPS [13]. Since then, a lot of attention has been paid to this methodology and other spectroscopic detection techniques have been implemented.

Classical ion-exchanger spectrophotometry can be defined as a non-chromatographic static technique consisting in the establishment of distribution equilibrium between the bulk of the sorbent and the analyte properly derivatized. After fulfillment of the steady-state regime of the heterogeneous batch reaction, which typically lasts 30 minutes, the sorbent phase is isolated from the solution by filtration or centrifugation. The colored beads are then transferred to a quartz cuvette as a suspension with a few microlitres of solution. Once the packing in the cuvette is uniformed, the spectrophotometric readings are carried out. In order to take into account the light-scattering caused by the sorbent itself and the walls of the cuvette, the preparation of a reference sorbent should be performed under identical experimental conditions [14].

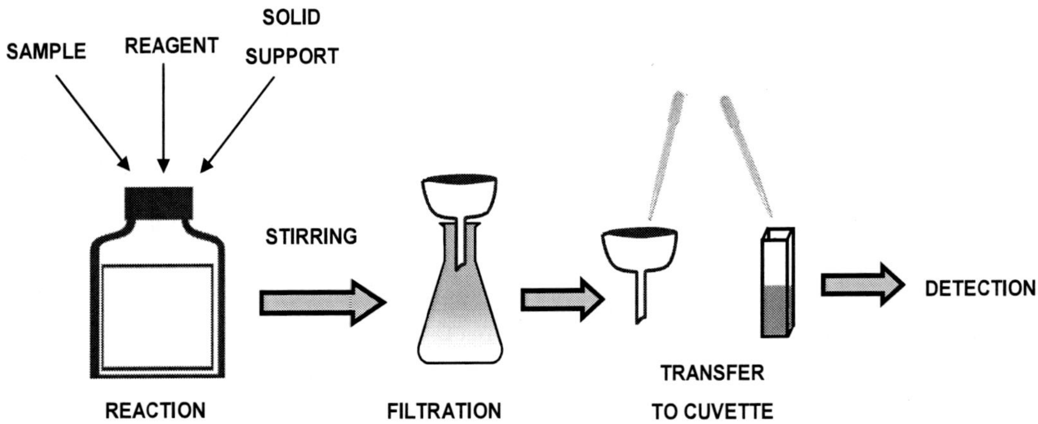

Figure 1. Solid-phase spectroscopy. Batch- mode.

The steps required for classical (or batch) SPS are detailed in Figure 1: a) the sample, reagents and solid support are put together and stirred, obtaining the reaction and adsorption of the complex on the solid beads; b) the solution is filtered to separate the solid support from the liquid solution; c) the solid support is transferred to a suitable cuvette; d) the detection is carried out.

Accordingly, the two main analytical features of SPS methods are its high sensitivity and selectivity, due to the preconcentration of the species of interest from a relatively large sample volume (typically 100-1500 mL) on a small amount of solid support, on which detection is performed directly with a non-destructive molecular spectroscopic detector. Hence, sensitivity is affected by the value of the ratio between sample volume and amount of solid support. Sensitivity values are several orders of magnitude greater than those ones from the respective conventional solution methods. Selectivity is also increased because all the potential interfering compounds that are not retained on the solid support stay in the liquid solution and, therefore, are separated from the analyte.

The most frequently used solid supports in SPS can be divided into three groups:

(a) Ion-exchanging polymers (Dowex styrene polymers or Sephadex dextran polymers). Usually, s*tyrene polymers*, constituted by a hydrophobic aromatic matrix, are discarded when working in UV region because of their very high background in this region. The skeleton of these resins (e.g. Dowex 1 and Dowex 50W) is jointed to different chemical groups. *Dextran polymers*, such as Sephadex absorbents, have been the most frequently used (e.g. Sephadex QAE A-25, Sephadex CM C-25 and Sephadex SP C-25), in which different functional groups can be introduced in order to produce cation- or anion- exchanger solid supports.
(b) Non-ionic polymeric adsorbents or neutral resins. Between these resins, also known like *macroporous polymers*, Amberlite XAD resin is the most used. One of the drawbacks of this type of supports is their high background in the UV and Visible regions. Although many fluorescent impurities due to the aromatic rings in its structure cause a serious background noise in the emission measurement, Amberlite XAD resin has usually been used in luminescence sensors.

(c) non-polar sorbents. The employment of a non-polar sorbent such as C_{18} bonded silica beads has allowed many determinations. However, an important drawback of this material, when compared to the other ones, is its lower selectivity because of the adsorptive nature of the retention process.

The sol-gel technology has also been introduced in the design of FTO. It provides a simple means to incorporate, at low temperature, organic and biological recognizing elements in a stable inert support. Typically, a sol is first formed by mixing a liquid alcoxide precursor, water, a co-solvent (usually ethanol or methanol) and a catalyst (acid or base) at room temperature. In this way, a porous gel network is obtained through continuous monomer hydrolysis and condensation reactions. Afterwards, gel aging and drying can be conducted in order to obtain densified solid matrices. During the hydrolysis steps, condensation or aging, the recognizing elements can be added and become entrapped in the support net, remaining sterically accessible to small analytes that diffuse into the pore network [15,16].

4. Flow Methodologies

The introduction of a solid support in conventional FIA systems was the beginning of flow-through optosensing although, as it has been previously commented, other flow methodologies have also been used. As a result, we will describe how FIA and MCFIA methodologies, the ones employed for the analysis of trace elements, work.

The basic components of any flow system include: propelling system, device/s for the introduction of solutions into the system, flow-through cell and detector.

In a FIA system, the basic components are a peristaltic pump to propel the sample and reagents, a series of plastic tubes to carry the solutions, injection valves to introduce constant volumes of sample and reagents in the system and the detector. A typical FIA manifold with spectrofluorimetric detection is shown in Figure 2.

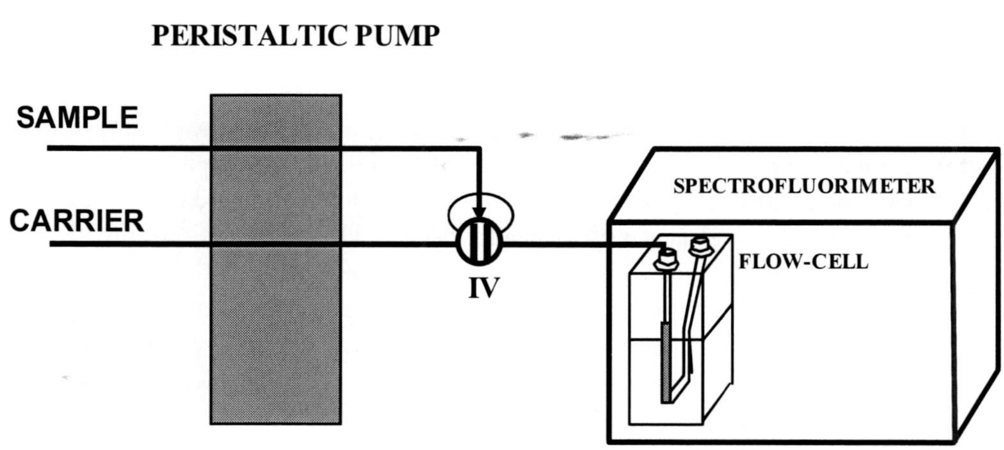

Figure 2. FIA manifold. IV: injection valve.

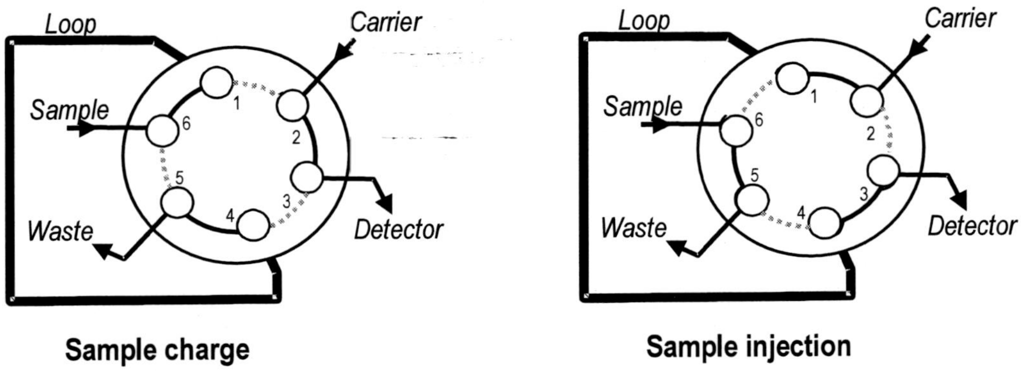

Figure 3. Procedure of a 6-way rotary valve.

The valves usually employed are six-way rotary valves, manually controlled. The sample charge and injection is performed by the movement of a mobile element between two fixed positions, being the sample volume selected by the length of the sample loop and the internal volume of the injection valve. The functioning of a rotary valve used as an injection valve is shown in Figure 3.

The main advantages of a FIA system are the rapidity, repeatability, versatility, automation and low consumption of reagents (when compared to batch methods).

In order to increase the automation and versatility of the FIA systems, SIA was introduced as a following generation in the development of flow injection technique. The main components of a SIA manifold are a multi-position selection valve and a bi-directional syringe pump, both automatically controlled by a computer. The use of the multi-position valve makes it unnecessary the use of additional valves, such as in conventional FIA systems, where several valves are needed. The main advantages of SIA are very simple manifolds, minimum solutions wasting, automation and robustness. However, FTO for the analysis of trace elements using SIA have not been developed up to date and research in this area would be interesting.

The most widely-used multicommutated methodologies are MCFIA, Multipumping Flow Systems (MPFS) and Multisyringe Flow Injection Analysis (MSFIA) [7]. MPFS are based on the use of individual solenoid micro-pumps, which act as both the propelling units and selection valves; these piston pumps are responsible for sample-reagent introduction and manifold commutation. MSFIA relies on the use of a multisyringe burette that is a multiple channel piston pump, driven by a single motor of a usual automatic burette; each syringe has a three-way solenoid valve placed at its head, which is used to select the direction of the liquid flow. Taking into account that MCFIA is the only methodology that has been coupled with SPS for trace elements determination [16-18], we will discuss it in more detail.

The key element in MCFIA is the solenoid valve, usually 3-way solenoid valves. Each valve acts as an independent switch and the whole system is automatically controlled by a computer. These systems are similar to an electronic circuit, with a variable number of active nodes, which present two different positions, "ON" and "OFF", allowing the effective control of sample and reagents dispersion and widening the scope of applications in flow analysis. Sample volume is selected by changing the flow-rate and the sample time insertion. Compared to conventional FIA, the insertion volume measurement is replaced with time measurement. Hence, the repeatability in multicommutated systems is associated not only with the stability of the peristaltic pump, but also with the precision in time measurement, obtaining the best repeatability when using electronic devices.

The main advantages of MCFIA are: (1) miniaturization of the manifolds due to the small size of the solenoid valves; (2) reduced sample and reagents consumption, as the solutions are recycled when required; (3) automation and high repeatability due to the absence of manual operations; (4) economy, as the solenoid valves are cheaper than the rotary valves; (5) versatility, due to the possibility of completely altering the flow of solutions and, therefore the whole system, without physical alterations, just changing the software procedure.

5. Flow-through Optosensing

The implementation of SPS in flow systems originated Flow Injection-Solid Phase Spectroscopy (FI-SPS) methodology. In this way, the advantages of SPS were added to those intrinsic ones from FIA. In addition to conventional FIA, MCFIA has also been coupled with SPS and applied to the determination of trace elements in different kinds of samples. In these FI-SPS systems the separation and retention of the species of interest on the solid phase takes place in the detection area itself and simultaneously with it. Therefore, these systems integrate (not only in space, but also in time) several analytical processes: a) separation and detection; b) reaction and detection; and c) reaction, separation and detection. The microzone of the solid phase where the signal is continuously monitored is integrated in the detector and surrounded by a continuous stream flowing through it. The sample plug is inserted in this stream and the radiation directly interacts with the solid surface integrated in the detection area. This detection principle, based on the interaction of the radiation with the solid surface integrated in the detector is called FTO. This approach is similar to a typical liquid chromatographic process with molecular spectroscopic detection, in which the detector displays the signal (in this case on the column itself) while the chromatographic separation is performed on the separation column (separation and detection occurring in the same place and at the same time, but at low pressure).

The use of a solid support in the detection zone (flow-through cell) is the key element in FTO. Solid supports used are packed in the flow-cell of a conventional non-destructive optical detector. The analytes or their reaction products are immobilised on them temporarily for sensing, so integrating retention and detection, sometimes even reaction.

A key aspect of FTO that differs from SPS performed in batch mode is the successive reutilization of the solid-phase microbeads placed in the flow cell for a large number of measurements. It involves regeneration of the solid support after each measurement, thus achieving reversible retention of the analyte to get the sensing zone ready for the next determination. This makes automatic flow SPS methodologies more environmental friendly than batch SPS, since a few mg of beads can be reutilized hundreds of times, therefore dramatically reducing the amount of solid support per determination. In addition, we need to highlight the single instrumentation, short analysis times and low cost per sample, all features making flow-through optosensing one methodology that can be performed in any laboratory and an interesting, cost-effective contributor to green analytical chemistry methodology.

Sometimes, solid supports packing microcolumns coupled on-line to the flow injection system have been used too. The employment of these microcolumns presents two main purposes: (a) interference removal by using a support suited to the sorption of the species to be removed; and/or (b) to get a multi-determination. The multi-determination can be accomplished by retaining selectively one of the analytes in the microcolumn (placed just before the flow cell) and eluting it later (after recording the signal from the first analyte) with an appropriate solution.

It should be taken into account that the solid support required for these FTO should meet some requirements: its particle size should be large enough to avoid overpressure in the system; it should be mechanically resistant to the continuous flow, chemically inert to the components of the solutions and compatible with the detection system used; and, finally, the retention/elution process should be quick enough.

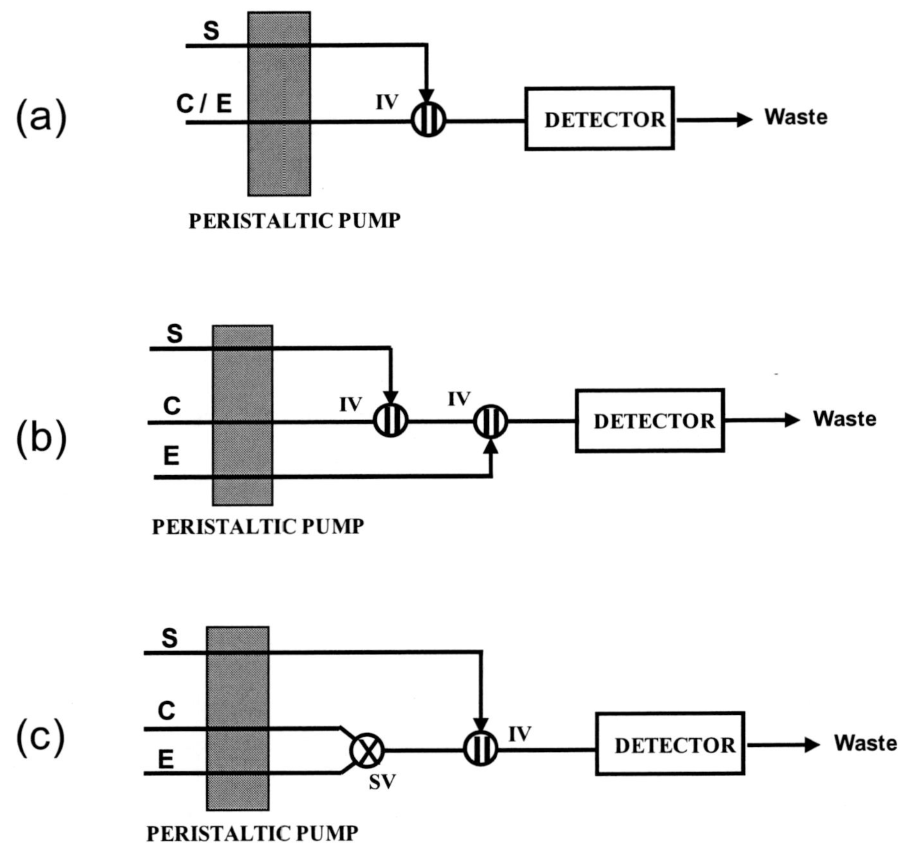

Figure 4. Different manifolds used for the regeneration of the solid support. C: carrier solution; E: eluting solution; S: sample solution; IV: injection valve; SV: selection valve.

In FTO, the solid support microbeads have to be regenerated after each sample determination, in order to prepare the sensing zone for the next analysis. The two main approaches employed for the regeneration have been traditionally:

(a) The carrier solution itself acts as the eluting agent. The regeneration process starts just as the tail end of the sample plug reaches the solid support in the flow-cell. In this case, a transitory signal will be developed in the detector. This is the simplest procedure to regenerate the solid sensing zone, and in turn, it allows high sampling frequencies (Figure 4a).

(b) By using an eluting solution, which is transported to the active solid zone after the maximum signal. This approach is used when the retention of the analyte on the solid microbeads is stronger and the carrier solution itself is not enough for the regeneration. This step can be performed via an injection valve (using a definite loop) (Figure 4b) or a selection valve (Figure 4c). It offers a lower sampling frequency although the obtained sensitivity is higher because the carrier solution does not elute the analyte from the microzone. In addition, it shortens the lifetime of the sensor when an ion exchanger is used as solid support, due to the successive swelling and compactation of the resin beads.

However, sometimes the retention of the target species on the solid support is so strong that none of the previously mentioned processes can regenerate the solid support. In these cases, a novel approach called Bead Injection Spectroscopy (BIS) was introduced. This methodology is employed only in cases of very strong retention of the analyte/s, in which the regeneration procedure is very difficult.

Bead Injection Spectroscopy

BIS is based on the renewal of the sensing zone by injection, transportation and trapping of a defined volume of fresh beads to the flow-cell before each analyte insertion, being the beads discarded to the waste once the measurement is completed [19,20]. The BIS concept, first proposed by Ruzicka et al., was introduced to use together with SIA [21]. It uses minute amounts of beads on which the reagent is adsorbed and the analyte is preconcentrated and monitored via optical fibers in specially designed (jet-ring-configured) flow cells. The bead suspension is pumped toward the cell and beads are retained while solution goes through a circular gap narrower than the diameter of the beads. After measurement is performed, beads are automatically discarded.

The disposal of the adsorbent beads and the injection of new sorbent beads into the flow-through cell are carried out after each analytical cycle. A bead injection (BI) cycle with its different steps is shown in Figure 5. The name "jet ring" cell comes from the configuration and operation of the bead retention cell made as part of the flow conduit by inserting a solid rod to block the beads in the assigned space.

The channel of the cell has an inner diameter slightly larger than the outer diameter of the blocking rod, forming the *O*-ring gap between the rod and the wall of the channel all around the rod. Only solutions are able to flow out while beads whose sizes are larger than the gap are packed due to the continuous incoming flow pushing the beads against the blocking rod. Used beads are discarded by the jet of solution reversely and rapidly flowing upstream.

Because the fresh adsorbent beads can always be used for both the retention and the optical sensing of analyte, this is applicable to target analytes which are strongly adsorbed on the solid phase and when the regeneration of the solid phase is difficult or the matrix species in samples are irreversibly adsorbed and lower the adsorption capacity of the adsorbent beads.

Some advantages in comparison to conventional FTO derive from its renewable nature, and they include: (1) suitability for implementing reagent-based assays without requiring full reversibility of the sorption/elution process; (2) long-term operation because sorbent compaction and clogging do not take place; and, (3) preservation of sorption behaviour because surface contamination and deactivation do not occur. However, two fundamental prerequisites to be fulfilled for the bead material to be explored in a renewable fashion are the feasibility of forming a stable suspension and reproducible manipulation throughout the automated system. Consequently, it is highly recommended ensuring bead-size homogeneity and spherical shape of the reagent-supporting entities in order to prevent compact settlement in the channels of the SI assembly. Hence, conventional, reversed-phase, chemically modified, silica-gel lumps are not really suitable for this purpose as a result of their irregular shape and size distribution. In spite of the advantages of these systems, some laboratories may not afford this relatively high-cost instrumentation, while a simpler FIA system implemented with the use of commercial flow cells can be as appropriate as SIA methodology is for this BIS approach. This BIS-FIA configuration has been developed by Ruedas Rama et. al., obtaining different FTO [19,20,22,23].

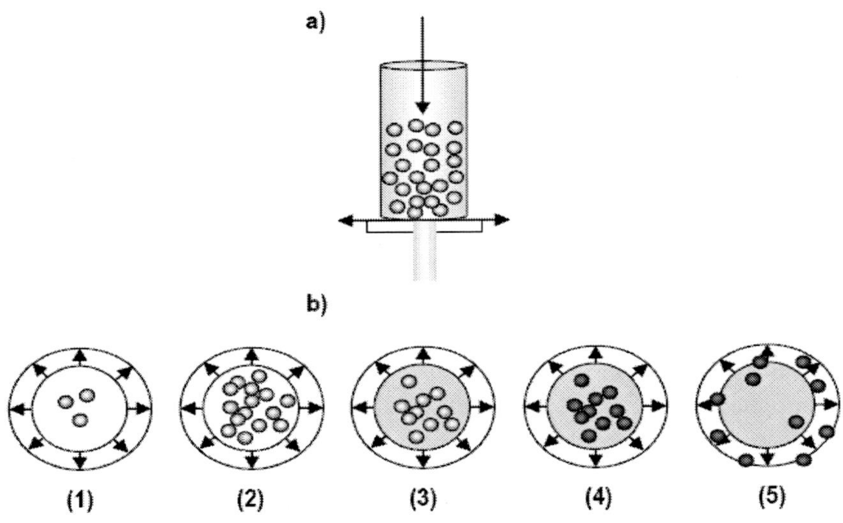

Figure 5. Jet ring cell diagram: (a) Front view; (b) Cross section (different steps).

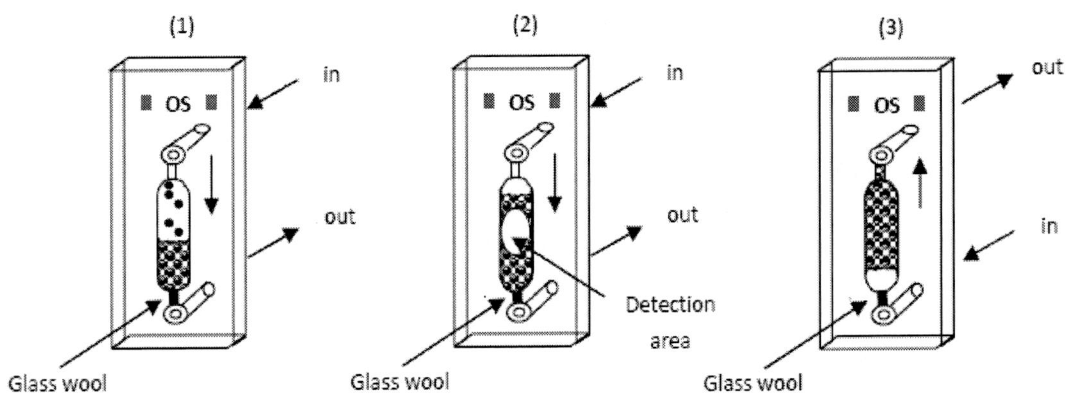

Figure 6. Different steps of the bead injection suspension into the flow cell: (1) Injection; (2) Detection; (3) Elimination.

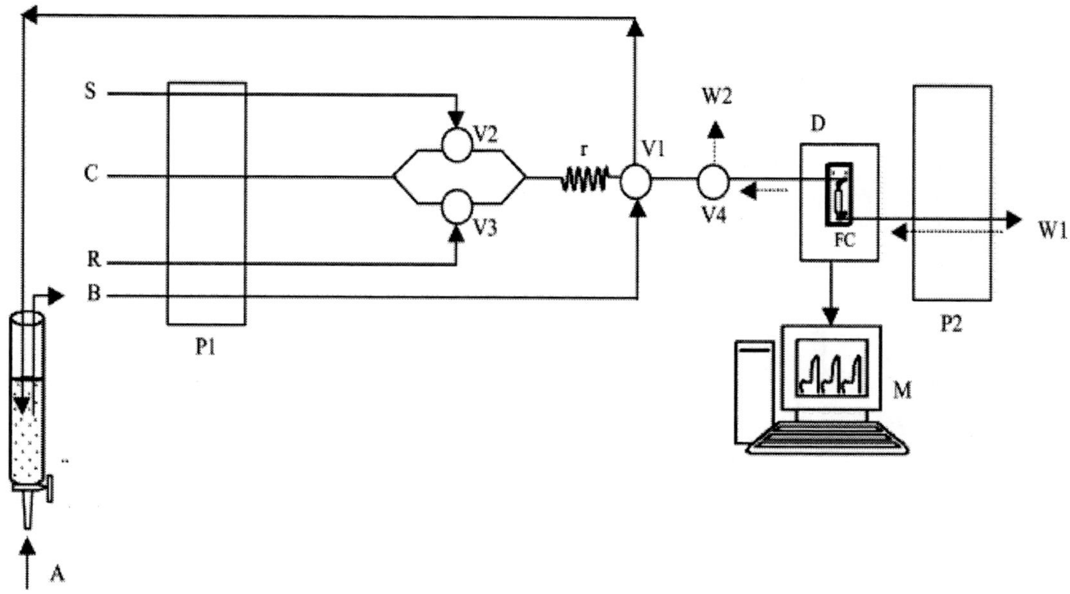

Figure 7. Manifold used in a BI system. A: air; S: sample; P1, P2: peristaltic pumps; C: carrier solution; R: reagent solution; B: bead suspension; V1, V2, V3: injection valves; V4: selection valve; r: reaction coil (50 cm); D: detector; FC: flow cell; W1, W2: waste; M: computer.

The procedure for each analysis (Figure 6) consists in the introduction of an exact volume of the bead suspension in the flow cell of a FIA system. Once the beads are trapped in the flow-cell and the baseline established, the sample is injected and the analytical signal obtained on the solid support. Finally, flow direction is reversed by means of a second peristaltic pump and the beads are automatically discarded and transported out of the system.

To demonstrate the utility of this approach, the system shown in Figure 7 was applied to the measurement of the absorbance of the complex formed between Fe(II) and ferrozine (sodium 3-(2-pyridyl)-5,6-diphenyl-1,2,4-triazine-4′-4′′-disulphonate) [19].

The procedure comprises five steps. (1) An exact volume of a bead suspension with the reagent immobilized on the surface is injected by means of the injection valve V1, loaded into the flow cell (Figure 6, step 1). (2) The beads are trapped in the cell and perfused with the carrier stream, and the baseline for the subsequent absorbance measurements is established. (3) The sample (Fe(II)) is injected via valve V2 and the analyte reaches the bead surface; it is chelated and the signal is continuously monitored (see Figure 6, step 2). (4) The reaction between reagent and analyte on the beads is accomplished. (5) Finally, beads are automatically discarded from the flow cell at the end of the assay cycle by reversal flow (see Figure 6, step 3). All these steps can be seen in the corresponding fiagram (Figure 8) which is registered along all the process.

Magnetic beads could also be used in the development of renewable surface optosensors. This innovative approach, which has not been used in trace element analysis, will widen the range of SIA or FIA applications. Since magnetic beads can be trapped in a flow system using a magnet, their retention in the flow cell as well as their discharge, can be easily accomplished with an (electro)magnet. Besides, magnetic beads are high-potential materials for a fully automated detection system because they have a large surface area and allow separation of target molecules from the reaction mixture.

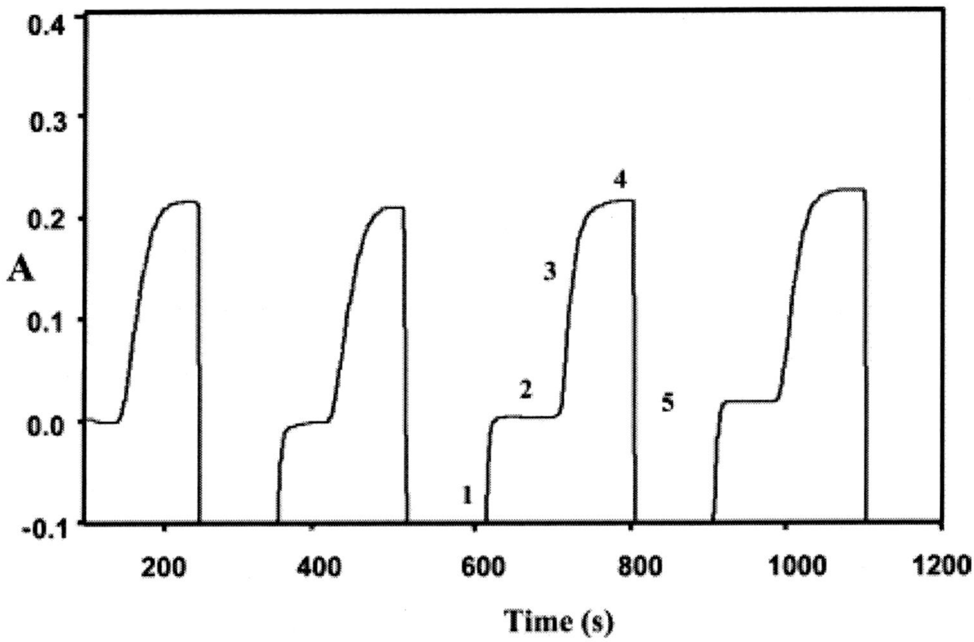

Figure 8. Fiagram obtained in a BI system.

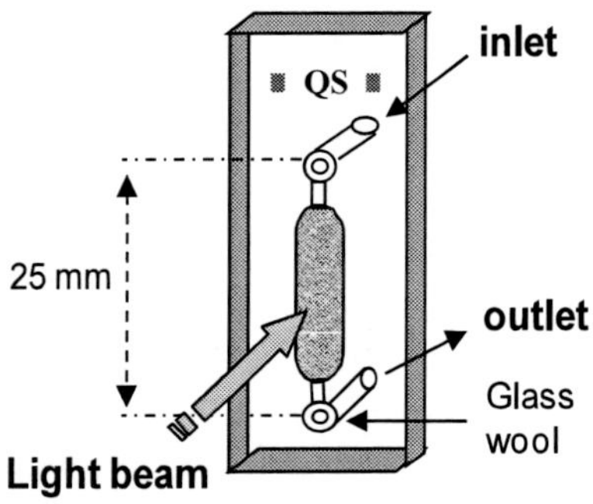

Flow-through Cell	Hellma 138 QS
Detection technique	UV-Visible spectroscopy
Light path	1.0 mm
Inner volume	50 µl

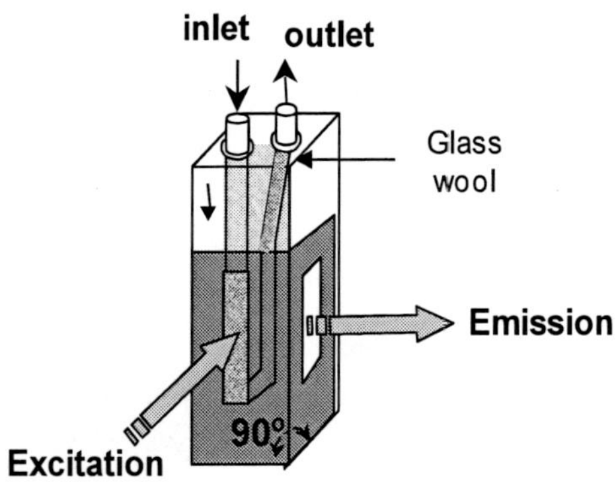

Flow-through Cell	Hellma 176.052 QS
Detection technique	Fluorescence Phosphorescence
Light path	1.5 mm
Inner volume	25 µl

Figure 9. Flow-through cells.

Flow-through Cells

Two main aims must be pursued when a suitable cell for the design of a flow-through sensor is being selected: (1) the concentration of the monitored product on the support in an area as small as possible of it; and, (2) the incident light beam must be focused to this area without loss of light to the surrounding zone. The best results are provided by cells with short path lengths (1–1.5 mm), which ensure compatibility between the system and the detector and prevent the species from lying outside the irradiated area. The most widely used flow-through cells are depicted in Figure 9.

The most appropriate commercial flow-through cell for spectrophotometric measurements in UV region is the Hellma 138-QS cell. The cell is blocked in the outlet with glass wool to prevent particle displacement by the carrier stream and the inlet is kept free. The level of the packing material into the cell should be the necessary one to fill it up to a sufficient height, thus permitting the light beam to pass completely through the solid layer. Higher levels would imply that the zone where the species of interest is sorbed would fall outside the detection area, obtaining lower and wider signals. On the other hand, lower levels would mean that the light beam would pass through the solution completely or partially and, consequently, a decrease in the signal would be obtained. Therefore, the top of the resin is kept as close as possible to the light beam, this latter being completely covered by the resin.

For fluorescence and phosphorescence measurements, the Hellma Model 176.052-QS flow-through cell with a light path of 1.5 mm is usually employed. In this case, in order to secure that the species of interest retained by the packing solid material is in the light path, the solid support level is maintained just some millimeters beyond the cell window. It is necessary to take into account that, when an eluting solution other than the carrier one is used to desorb the species of interest from the solid support, ion-exchange resins suffer alternately swelling and compaction, therefore altering the level of the support in the cell and the baseline. This is due to the different chemical nature and concentration of the carrier and the eluting solutions. Hence, in this case, and for preventing compactions lowering the support level below the light beam, the flow cell must be filled up passing the eluting solution through it. In every case, it is necessary to condition the solid support by passing the carrier solution through it for a few minutes.

6. Detection Techniques

Different spectroscopic detection techniques have been used in the development of FTO and they will be briefly described below.

Absorptiometry

UV-Visible spectroscopy has been traditionally the most frequently used detection technique in analytical laboratories due to its high flexibility for adaptation to a wide variety of analytical problems and the low-cost of equipments. However, not many analytical methods rely on the intrinsic absorbance of the analytes, as their intrinsic light absorption occurs at the UV region, where other co-existing compounds also absorb, interfering in the analytical measurements. As a result, UV region does not usually present the required selectivity for conventional spectrophotometric analysis. However, the implementation of the active solid support in spectrophotometric flow systems strongly enhances selectivity, excluding from the detection area all those species that cannot be retained in the experimental conditions.

Different analytes have been determined simply by measuring their native absorbance. In all the described spectrophotometric methods, derivatization reaction between the trace element and a chromogenic agent is required before the measurement. In general, two approaches are available: chelation before the solid support or chelation on the same solid support. However, the previous loading of the chromogenic agent on commercial solid supports [19,22] or its incorporation in sol-gel films during their preparation [16-18] has been the chosen option. In this way, a homogeneous distribution of the reagent on the solid support is achieved, increasing the repeatability and sensitivity of the system.

Luminescence

Luminescence detection techniques have gained considerable ground in analytical chemistry due to the high sensitivity and selectivity that can be achieved. However, not all luminescence techniques are selective enough, and different strategies can be employed in order to overcome some limitations, such as spectral overlapping, scattered light, quenching processes or background signal. The techniques employed in FTO have been fluorescence, phosphorescence, chemiluminscence, and lanthanide sensitized-luminiscence, being only the first two employed for the determination of trace element.

Fluorescence

In the case of fluorescent light measurement emitted from the solid phase in the cell, utilizing a longer light path cell is not effective because the deeper part of the solid particles in the cell cannot be irradiated by the excitation light beam. To increase the excitation light beam intensity is also unsuitable for sensitivity enhancement because the scattered or reflected excitation light enters the emission light detector even if it is located perpendicular to the excitation light beam, which results in a serious emission background error. However, because the sensitivity of fluorometry is inherently higher than that of absorptiometry, a solid phase light path of 1 to 1.5 mm is sufficient for trace analysis, and the measurement of emitted light from the solid phase can be easily accomplished in a way similar to that with solution spectrometry. Compared with solid phase absorptiometry, the applicability of solid phase spectrofluorometry, especially for determination of metal ions, has been limited because of the lower number of fluorogenic reagents selective for a specific metal ion.

In fluorometric optosensors, both the previous loading of the reagent on the solid support [20,23] and the on-line formation of the chelate before the flow-cell [24,25] strategies have been used. When the reaction takes place before the cell, sometimes it is necessary to use a reaction coil to fulfill the chelation [24]. This strategy lowers the sensitivity of the system and decreases the sample throughput.

Phosphorescence

Under specific conditions, phosphorescence from chemical components can be observed even at room temperature. This long-lived emission is called room temperature phosphorescence (RTP).

When comparing RTP methods with fluorometric ones, RTP detection presents some advantages over fluorometry: the short-lived background luminescence or scattered light can be easily discriminated because of the long-lived emission of the analyte species, and very good separation of the maximum wavelength of phosphorescence from that of the excitation spectra due to Stoke's shifts is achieved.

FTO with RTP detection are similar to those ones with fluorometric detection, except for the optimization of the gate and delay times of the detector, in order to isolate the RTP signal. To observe RTP in solution (or from the solid phase in the flow-through cell), the existence of a high concentration of heavy atom salts such as thallium nitrate or iodide, which are thought to play an important role in facilitating the energy transition from excitation singlet state to excitation triplet state of the phosphorescent species, is very effective. Oxygen scavengers such as sodium sulfite have to be added too.

In spite of the advantages of RTP over fluorometry, not many reports on the application of RTP have been published [26,27]. In both papers, derivatives from quinolinesulphonic acid were employed to complex the inorganic elements. In one of them [27], the chelates formed between lead (II) and the reagents 8-hydroxy-5-quinolinesulphonic acid, 8-hydroxy-7-quinolinesulphonic acid and 8-hydroxy-7-iodo-5-quinolinesulphonic acid exhibited strong RTP if retained on the surface of anion exchange resin beads. The chelation was formed on-line and the solid support was easily regenerated using a fixed volume of hydrochloride acid. This method was applied to the determination of lead in sea water at few ng mL^{-1} levels.

7. APPLICATIONS

The main applications of FTO in the field of trace elements are described in Tables 2 and 3. The first one corresponds to the spectrophotometric methods (seven) and the second one to the luminescence methods (eight using fluorescence and two using RTP). As can be seen, the analyzed samples belong to very different areas of research, such as clinical, environmental or alimentary.

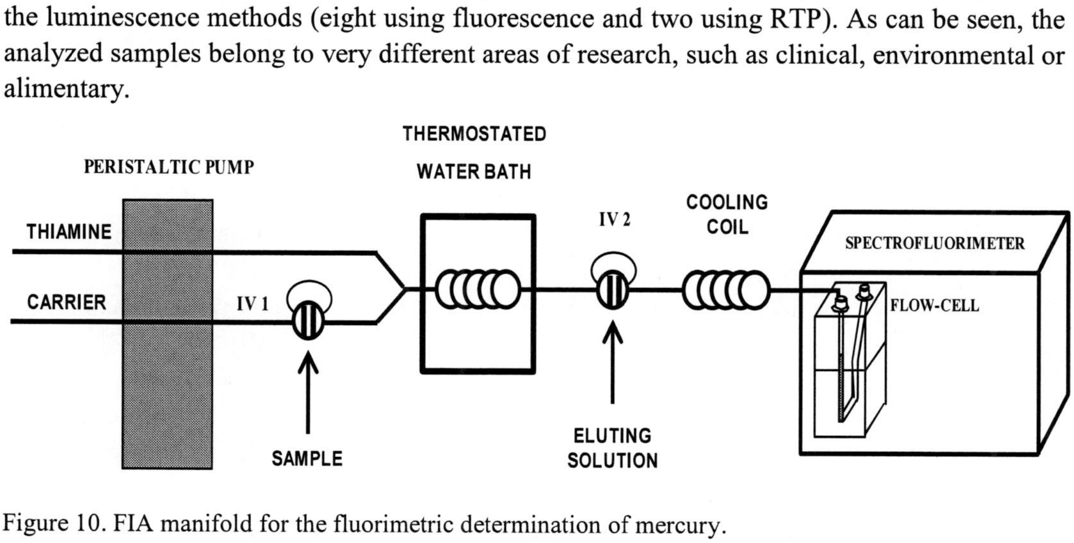

Figure 10. FIA manifold for the fluorimetric determination of mercury.

The analysis of environmental samples, such as soil [22], river waters [23,28] or sea water [27] has focused not only in essential trace elements, such as iron [19,29] or vanadium [20], but also in the analysis of toxic elements. For instance, there is a great demand for the determination of lead in environmental samples because this metal is now widespread, contaminating virtually the whole biosphere, and its determination in sea water has been performed using RTP detection [27]. Chromium (Cr) is mostly present in two different oxidation states, Cr (III) and Cr (VI). Cr (III) is an essential trace element. However, Cr (VI) is very toxic to organisms and it has been demonstrated that its compounds are carcinogenic to humans. An on-line oxidation method for determining chromium in waters was developed using a flow electrolysis cell, using diphenylcarbazide as the chromogenic agent. This system was applied to the total Cr determination together with the specific determination of Cr (III) and Cr (VI) at trace levels in river water samples [28].

A fluorescence optosensor was described for the determination of mercury (II) in water samples [30]. The FIA method was based on the immobilization on a non-ionic-exchanger solid support of the thiochrome formed by the oxidation of thiamine with mercury (Figure 10). Sample or standard solutions were injected via IV 1 into the carrier stream.

This stream was mixed with the thiamine solution in a T-piece and passed through the reaction coil (immersed in a thermostatted water bath to increase the kinetics of the thiamine oxidation). A cooling coil was used to decrease the flow temperature to the room temperature. Finally, the thiochrome formed went through the flow cell where it was retained on the packed resin and the fluorescence signal was recorded. After that, an eluting solution was inserted via IV 2 to regenerate the solid beads.

Vanadium is known in the marine environment mainly due to its high concentration in tunicates, in which it is present at levels about seven times greater than those in seawater. In spite of the fact that studies of vanadium in molluscs are rather scarce, there have been some reports pointing to the vanadium enrichment of marine organisms living close to vanadium contamination sources, such as the titanium dioxide industry, harbours, and petrochemical plants, which would suggest that molluscs also accumulate vanadium [33]. Therefore, the determination of this metal in mussels may be a good control parameter of the contamination levels. Using a BI system, the determination of vanadium could be accomplished in mussel tissues [20]. In the mentioned paper, a homogeneous bead suspension of an anionic-exchanger resin (previously loaded with the fluorogenic reagent Alizarin Red S) was injected to fill the flow cell. Then, the sample containing vanadium (V) was injected into the carrier and reacted with the immobilized reagent on the active solid support placed in the flow cell to form a fluorescent chelate. The complex was so strongly retained on the beads that the regeneration of the solid support became extraordinarily difficult and it had to be changed after each sample determination. The manifold required was similar to that one depicted in Figure 7.

It is important to mention that the determination of more than one element can also be performed by modifying the flow system. In this way, aluminium (III) and beryllium (II) could be fluorometrically determined using a BIS–FIA system [23]. The sensor was based on the alternate use of two carrier solutions. The flow-cell was filled by injecting a homogeneous bead suspension of an appropriate solid support (Sephadex QAE A-25) previously loaded with the fluorogenic reagent morin. By using the first carrier solution, only the reaction between aluminium and morin was accomplished when the sample was inserted, obtaining the corresponding signal. After that, the carrier was changed and the sample solution inserted once more. This time, only beryllium reacted with morin. At the end of the analysis, beads were automatically discarded from the flow cell, by reversing the flow, and transported out of the system. Hence, both metals could be determined using the same flow system, by just changing the carrier solution. This multisensor was applied to the determination of these elements in tap and river waters. The same authors developed a similar system for the resolution of mixtures of copper (II) and zinc (II) using spectrophotometric detection [22]. In this case, the chromogenic agent was Zincon and two different carrier solutions were also used in order to selectively form the appropriate complex.

Trace elements have been determined not only in environmental waters, but also in drinking waters. In this way, FTO have been developed for the determination of aluminum, cadmium and zinc [24,25,31]. Zinc is an essential trace element. In the case of aluminum, although low concentrations are innocuous for human health, high concentrations can lead to some undesirable effects. Cadmium, however, is a toxic metal and its maximum permissible level in drinking water is 5 μg L^{-1}. All these optosensors were developed using a similar approach: an on-line reaction between the target element and a fluorogenic reagent was carried out before the flow-through cell and the formed chelate was then retained on the solid support. In all cases, satisfactory sensitivity and selectivity were observed for the determination of these elements in drinking water at the required levels.

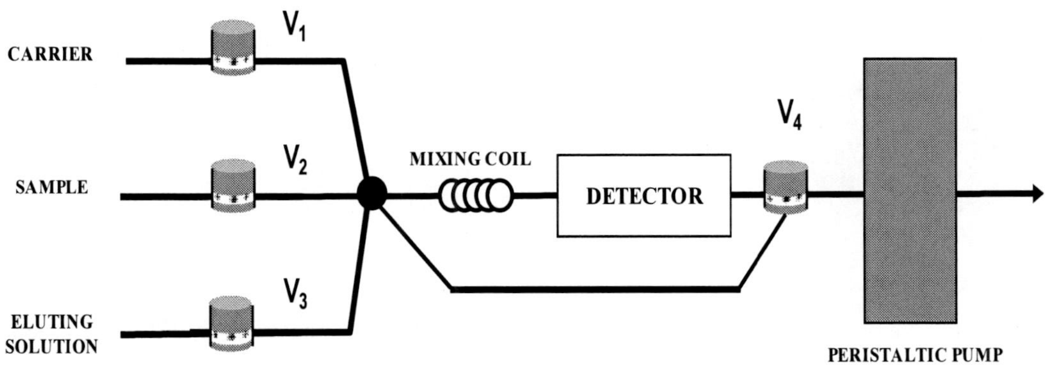

Figure 11. MCFIA manifold for the determination of bismuth, copper or zinc.

Finally, different FTO have also been developed in clinical analysis. Some of the previous systems were applied not only to the mentioned applications, but also to the analysis of those trace elements in pharmaceutical preparations [19,22] or physiological samples [20]. As it has been described, in addition to toxic elements, trace essential elements can also be toxic for human health when surpassing certain concentrations. Hence, the selective and sensitive analysis of these elements in clinical samples is mandatory. All the applications previously described were carried out using FIA or BIS-FIA.

However, as mentioned in the "Flow methodologies" section, MCFIA has also been applied to the analysis of trace elements. Three different optosensors, using sol-gel as sensing support, have been described for the determination of bismuth (III) [16], copper (II) [18] and zinc (II) [17] in pharmaceutical preparations or urine samples (Figure 11). In these sensors, a colorimetric organic compound was entrapped in the sol-gel inert support in order to form a complex with the analyte when arriving to the flow-cell (placed in the detector). The substitution of solutions accessing the system through valves V_1, V_2 and V_3 was made possible by simultaneous activation of each one of these valves together with valve V_4. The sample flow through the mixing coil was achieved by simultaneous activation of valves V_1 and V_2, using the binary sampling approach (introducing alternative small plugs of each solution to obtain a homogeneous mixing). Once the absorbance of the complex was recorded, the solid support was easily regenerated by activating valve V_3 and V_4.

Table 2. Applications of FTO using spectrophotometric detection

Analyte	Metho-dology	Solid support	Matrix	Pre-treatment	DL ($\mu g\ mL^{-1}$)	Reference
Chro-mium	FIA	Muromac 50W-X2	Waters	Reaction with diphenylcarbazide	0.014	[28]
Copper (II) Zinc (II)	BIS-FIA	QAE A-25	Waters, pharmaceuticals, soil, human hair	Reaction with 2-carboxyl-2-hydroxy-5-sulfoformazylbenzene (Zincon)	29 40	[22]
Iron (II)	FIA	C_{18} silica gel	Waters	Reaction with 1-(2-thiazolylazo)-2-naphthol	15	[29]
Iron	BIS-FIA	QAE A-25	Pharmaceuticals; waters; wine	Reaction with ferrozine	24	[19]
Bismuth (III)	MCFIA	Sol-gel	Pharmaceuticals	Complexation with xylenol orange	7	[16]
Copper (II)	MCFIA	Sol-gel	Urine	Complexation with 4-(2-pyridylazo)resorcionol	3	[18]
Zinc (II)	MCFIA	Sol-gel	Pharmaceuticals	Complexation with 4-(2-pyridylazo)resorcionol	2	[17]

D.L.: detection limit.

8. Future Trends

In this chapter, we have described the use of FTO for the determination of trace elements in different kinds of samples, especially waters and pharmaceuticals. Most of the mentioned methods of analysis are rapid, simple and versatile, showing high sensitivity and selectivity. However, further research in the following points would be interesting to improve the suitability of FTO in trace elements determination:

- Molecular imprinted polymers could be employed as solid supports in FTO. They are synthesized using the target compound as the template, therefore obtaining high specificity for the desired analyte.
- The use of magnetic beads as solid support would widen the scope of applications of BIS. Since these beads can be trapped in a flow system using a magnet, their retention in the flow cell as well as their discharge, can be easily accomplished with an (electro)magnet.
- RTP detection technique has been scarcely used and, as it was previously mentioned, it presents interesting advantages when compared with other detection techniques.
- Only a small number of FTO were developed using multicommutation. The introduction of alternative flow methodologies, such as MPFS or MSFIA, could be useful in order to develop portable instruments for *in situ* measurements.

Table 3. Applications of FTO using luminescence detection

Analyte	Methodology	Solid support	Matrix	Pre-treatment	DL ($\mu g\ mL^{-1}$)	Detection	Reference
Aluminium (III) Berillium (II)	BIS-FIA	QAE A-25	Waters	Reaction with morin	10 24	Fluorescence	[23]
Lead (II)	FIA	Dowex 1x2-200	Sea water	Reaction with derivatives of quinolinesulphonic acid	0.1	RTP	[27]
Mercury (II)	FIA	Amberlite XAD-4	Waters	Reaction with thiamine	3	Fluorescence	[30]
Vanadium (V)	BIS-FIA	QAE A-25	Waters, serum, urine, mussels	Reaction with Alizarin Red S	450	Fluorescence	[20]
Aluminium (III)	FIA	C_{18} silica gel	Drinking waters	Reaction with chromotropic acid	2.6	Fluorescence	[24]
Cadmium (II)	FIA	QAE A-25	Drinking waters	Reaction with 8-hydroxyquinoline-5-sulfonic acid	0.48	Fluorescence	[25]
Zinc (II)	FIA	C_{18} silica gel	Drinking waters	Reaction with p-(tosylamino)quinoline	0.9	Fluorescence	[31]
Aluminium (III)	FIA	Chelex 100	---	Reaction with quinolin-8-ol-5-sulfonate	0.08	Fluorescence	[32]
Iodide	FIA	Dowex 1x2-200	---	Reaction with chelate aluminium-quinolin-8-ol-5-sulphonic acid	10000 5000	RTP Fluorescence	[26]

D.L.: detection limit.

The main handicap of these sensors is the impossibility to perform multi-residue analysis, as more than two trace elements have not been determined up to date in the same analytical method. Taking into account that complex samples can be analyzed using these FTO, the main trends would be: (1) design of miniaturized instruments for in-situ measurements; (2) further investigation in order to determine several analytes in the same analysis.

REFERENCES

[1] Reinhold, J.G. (1975). Trace elements: a selective survey. *Clinical Chemistry*, 21, 476-500.

[2] Nancy, L. (2011). The use of atomic spectroscopy in the pharmaceutical industry for the determination of trace elements in pharmaceuticals. *Journal of Pharmaceutical and Biomedical Analysis*, 55, 653-661.

[3] Taylor, A., Branch, S., Day, M.P., Patriarca, M. and White, M. (2011). Atomic spectrometry update. Clinical and biological materials, foods and beverages. *Journal of Analytical Atomic Spectrometry*, 26, 653-692.

[4] Evans, E.H., Day, J.A., Palmer, C.D. and Smith, C.M.M. (2011). Advances in atomic spectrometry and related techniques. *Journal of Analytical Atomic Spectrometry*, 26, 1115-1141.

[5] Capitán-Vallvey, L.F. and Palma, A.J. (2011). Recent developments in handheld and portable optosensing-A review. *Analytica Chimica Acta*, 696, 27-46.

[6] Ruzicka, J. and Hansen, E.H. (1975). Flow injection analyses. Part I. A new concept of fast continuous flow analysis. *Analytica Chimica Acta*, 78, 145-157.

[7] Cerdá, V., Forteza, R. and Estela, J.M. (2007). Potential of multisyringe flow-based multicommutated systems. *Analytica Chimica Acta*, 600, 35-45.

[8] Ruzicka, J. and Marshall, G.D. (1990). Sequential injection: A new concept for chemical sensors, process analysis and laboratory assays. *Analytica Chimica Acta*, 237, 329-343.

[9] Llorent-Martínez, E.J., Ortega-Barrales, P., Fernández-de Córdova, M.L. and Ruiz-Medina, A. (2010). Multicommutation in flow systems: A useful tool for pharmaceutical and clinical analysis. *Current Pharmaceutical Analysis*, 6, 53-65.

[10] Wang, J. and Hansen, E.H. (2003). Sequential injection lab-on-valve: The third generation of flow injection analysis. *TrAC - Trends in Analytical Chemistry*, 22, 225-231.

[11] Fraga, C.G. (2005). Relevance, essentiality and toxicity of trace elements in human health. *Molecular Aspects of Medicine*, 26, 235-244.

[12] Goldhaber, S.B. (2003). Trace element risk assessment: Essentiality vs. toxicity. *Regulatory Toxicology and Pharmacology*, 38, 232-242.

[13] Yoshimura, K., Waki, H. and Ohashi, S. (1976). Ion-exchanger colorimetry-I. Micro determination of chromium, iron, copper and cobalt in water. *Talanta*, 23, 449-454.

[14] Miró, M. and Frenzel, W. (2004). Flow-through sorptive preconcentration with direct optosensing at solid surfaces for trace-ion analysis. *TrAC - Trends in Analytical Chemistry*, 23, 11-20.

[15] Jerónimo, P.C.A., Araújo, A.N. and Conceiçao, B.S.M. (2007). Optical sensors and biosensors based on sol-gel films. *Talanta*, 72, 13-27.

[16] Jerónimo, P.C.A., Araújo, A.N., Montenegro, M.C.B.S., Satinský, D. and Solich, P. (2004). Colorimetric bismuth determination in pharmaceuticals using a xylenol orange sol-gel sensor coupled to a multicommutated flow system. *Analytica Chimica Acta*, 504, 235-241.

[17] Jerónimo, P.C.A., Araújo, A.N. and Montenegro, M.C.B.S. (2004). Development of a sol-gel optical sensor for analysis of zinc in pharmaceuticals. *Sensors and Actuators, B: Chemical*, 103, 169-177.

[18] Jerónimo, P.C.A., Araújo, A.N., Montenegro, M.C.B.S., Pasquini, C. and Raimundo, J. (2004). Direct determination of copper in urine using a sol-gel optical sensor coupled to a multicommutated flow system. *Analytical and Bioanalytical Chemistry*, 380, 108-114.

[19] Ruedas Rama, M.J., Ruiz Medina, A. and Molina Díaz, A. (2003). Bead injection spectroscopic flow-through renewable surface sensors with commercial flow cells as an alternative to reusable flow-through sensors. *Analytica Chimica Acta*, 482, 209-217.

[20] Ruedas Rama, M.J., Ruiz Medina, A. and Molina Díaz, A. (2005). A flow-injection renewable surface sensor for the fluorimetric determination of vanadium(V) with Alizarin Red S. *Talanta*, 66, 1333-1339.

[21] Ruzicka, J. and Scampavia, L. (1999). From flow injection to bead injection. *Analytical Chemistry*, 71, 257A-263A.

[22] Ruedas-Rama, M.J., Ruiz-Medina, A. and Molina-Díaz, A. (2005). Resolution of biparametric mixtures using bead injection spectroscopic flow-through renewable surface sensors. *Analytical Sciences*, 21, 1079-1084.

[23] Ruedas Rama, M.J., Ruiz Medina, A. and Molina Díaz, A. (2004). Implementation of flow-through multi-sensors with bead injection spectroscopy: Fluorimetric renewable surface biparameter sensor for determination of berillium and aluminum. *Talanta*, 62, 879-886.

[24] García-Reyes, J.F., Ortega-Barrales, P. and Molina-Díaz, A. (2005). Development of a solid surface fluorescence-based sensing system for aluminium monitoring in drinking water. *Talanta*, 65, 1203-1208.

[25] García-Reyes, J.F., Ortega-Barrales, P. and Molina-Díaz, A. (2006). Sensing of trace amounts of cadmium in drinking water using a single fluorescence-based optosensor. *Microchemical Journal*, 82, 94-99.

[26] Liu, Y.M., Garcia, R.P., Diaz Garcia, M.E. and Sanz-Medel, A. (1991). Room-temperature luminescence optosensing based on immobilized metal chelates: Application to iodide determination. *Analytica Chimica Acta*, 255, 245-251.

[27] San Vicente De La Riva, B., Costa-Fernández, J.M., Pereiro, R. and Sanz-Medel, A. (1999). Flow-through room temperature phosphorescence optosensing for the determination of lead in sea water. *Analytica Chimica Acta*, 395, 1-9.

[28] Matsuoka, S., Nakatsu, Y., Takehara, K., Saputro, S. and Yoshimura, K. (2006). On-line electrochemical oxidation of Cr(III) for the determination of total Cr by flow injection-solid phase spectrophotometry. *Analytical Sciences*, 22, 1519-1524.

[29] Teixeira, L.S.G. and Rocha, F.R.P. (2007). A green analytical procedure for sensitive and selective determination of iron in water samples by flow-injection solid-phase spectrophotometry. *Talanta*, 71, 1507-1511.

[30] Segura-Carretero, A., Costa-Fernández, J.M., Pereiro, R. and Sanz-Medel, A. (1999). Low-level mercury determination with thiamine by fluorescence optosensing. *Talanta*, 49, 907-913.

[31] García-Reyes, J.F., Ortega-Barrales, P. and Molina-Díaz, A. (2007). Flow-through fluorescence-based optosensor for the screening of zinc in drinking water. *Analytical Sciences*, 23, 1179-1183.

[32] Gong, Z. and Zhang, Z. (1996). An optosensor based on the fluorescence of metal complexes adsorbed on Chelex 100. *Analytica Chimica Acta*, 325, 201-204.

[33] Saavedra, Y., Fernández, P. and González, A. (2004). Determination of vanadium in mussels by electrothermal atomic absorption spectrometry without chemical modifiers. *Analytical and Bioanalytical Chemistry*, 379, 72-76.

In: Trace Elements ISBN: 978-1-62081-376-8
Editors: Diego A. De Leon and Paloma Raquel Aragon

Chapter 3

TRACE METALS IN FRUIT AND VEGETABLE AND THEIR EFFECTS ON HUMAN HEALTH

Stefania Papa, Giovanni Bartoli and Antonietta Fioretto
II University of Naples, Department of Life Science, Caserta, Italy

ABSTRACT

Trace metals occur naturally in the earth's crust, but human activities have greatly increased the mobilization and bioavailability of metals and, therefore, the chance of exposure to harmful concentrations. In fact, they along with other pollutants are discharged in the environment through industrial activity, car exhaust, heavy-duty electric power generators, refuse burning and use of pesticides in agriculture etc.

Natural baseline levels of trace metals in air, soil, rivers, lakes, and oceans are usually low, but in certain forms, and at sufficiently high concentrations, trace metals can be toxic to organisms.

Trace elements play an important role in biological and biochemical reactions in the living cells. They are stable and cannot be metabolized by the organisms, so, through the food, they tend to bio-accumulate. If the accumulation rate is higher than the detoxification rate, they can reach toxic concentration levels. Toxic metals in the human body can "replacing" essential nutrients, especially if deficient; for example Pb can replace Ca, Al can replace Mg, etc.. The substitution allows the enzyme systems to continue its function, but, if the intoxication degree exceed the tolerable limits, serious diseases take over.

The major source of trace metals for man is the food, in particular, fruit and vegetables that are essential components of the diet, by contributing protein, vitamins, iron, calcium and other nutrients in short supply. In recent years their consumption is gradually increasing, particularly in the urban community. However, because of the increased environmental pollution, fruit and vegetables can also contain trace metals over a wide range of concentrations. That because the plants take off them from soil as well as from air particulate. In addition the contamination of fruit and vegetable products may be due also through irrigation with contaminated water, the use of fertilizers and metal-based pesticides, industrial emissions, transportation, the harvesting process, storage and/or at the point of sale.

To monitor metal contents in these products is very important for crop yield technology, food nutrition and health impact. Many studies show that the trace metals at relatively high concentration may disturb the normal physiology (the normal functions of central nervous system, liver, lungs, heart, kidney and brain, produce hypertension, abdominal pain, skin eruptions, intestinal ulcer) until to produce different types of cancer.

INTRODUCTION

The trace metals (TMs) are among the most dangerous pollutants for the environment and human health. They occur naturally in the environment but, in the last decades, because the rapid development of industry, agriculture, traffic and transport, mining activities, their mobilization and bioavailability have been greatly increased. They are not biodegradable and have a long biological half-life, so usually bio-accumulate (Radwan and Salama, 2006).

The trace metals may enter the human body through inhalation of dust, consumption of contaminated drinking water, direct ingestion of soil and consumption of plant foods grown up in contaminated soils (Charya et al., 2008). Nevertheless, the ingestion of contaminated foods is considered one of the main streets through which TMs enter the human body. Some TMs are essential for the functioning of its biological systems, such as zinc and copper which act as a structural and functional constituents of proteins, but many others are pollutants, such as cadmium, lead, chromium, nickel and vanadium. In general, an excessive retention of all TMs in the environment systematically induce toxic effects and genomic instability and, in addition, cadmium, lead and mercury affect the endocrine system.

TMs contamination of agricultural soils, resulting from mining activities, agricultural practices, a part from natural phenomena of volcanism, raises environmental problems in the long term, because the anthropogenic sources of these elements strongly affect their speciation and bioavailability (Muchuweti et al., 2006).

In general, there are many factors that influence the concentration of trace metals on and in plants. These factors include: a) climate; b) atmospheric deposition; c) soil nature; d) maturity of the plant; e) agricultural systems; f) use of fertilizers; g) irrigation with contaminated water; etc. Moreover, the soil nature is considered the main factor involved in the transfer of the TMs in plants destined to human consumption.

Fruits and vegetables are an important part of the human diet because they contain a low energy content, but are rich in nutrients such as protein, vitamins, essential macronutrients, trace elements and fibers. The daily consumption of fruits and vegetables (300g/day) helps prevent cardiovascular disease and certain cancers (WHO, 2003). For this reason it is important to monitor the content of trace elements in plant foods (Ekholm et al., 2007).

At the basis of the above, programs have been initiated to evaluate and monitor the content of trace metals through the food chain in most developed countries (Xiu-Zhen et al., 2009).

Sources of Heavy Metals

Heavy metals are a group of elements with a mass density greater than 4.5 g/cm^3, which tend to release electrons in chemical reactions and form simple cations. In the solid and liquid states they are characterized by good heat and electrical conductivity, and are glossy and opaque. They have high melting and boiling points and are malleable with usually monoatomic pairs. The metals classified as heavy metals include: Cu, Co, Cr, Cd, Fe, Zn, Pb, Sn, Hg, Mn, Ni, Mo, V, W. Within the group of heavy metals it can distinguish both the elements essential for living organisms (microelements) and the elements whose physiological role is unknown, and thus they are “inactive” towards plants, animals and people. The metals serving as microelements in living organisms usually occur in trace amounts that are precisely defined for each species. Both their deficiency and excess badly affect living organisms (Figure 1).

The strongest toxic properties are characteristic for inorganic metals compounds, which dissociate well and are easily soluble, because they can easily penetrate through cell membranes and get into internal organs.

The circulation and migration of metals in the natural environment are mainly related to such processes as rock decay, volcano eruptions, evaporation of oceans, forest fires and soil formation processes. The sources of anthropogenic contamination or pollution of the environment by heavy metals include different branches of industry, the power industry, transport, municipal waste management, waste dumping sites, fertilizers and waste used to fertilize soil. The heavy metals from these sources are dispersed in the environment and they contaminate soil, water and air. They also (directly or indirectly through plants) get into human and animal bodies.

A general scheme of the possible inputs of natural and anthropogenic metals in water and soil is shown in Figure 2. Generally, it can be observed that more a soil is developed, less is the influence of the parent rock on its trace metals content (Whedepohl, 1969).

Besides, semimetals (metalloids), such as arsenic, are the elements that have properties intermediate between those of metals and nonmetals. The metalloids have a number of properties characteristic for metals, such as a glossy surface in the solid state and high melting points. Although their electric and heat conductivity is much lower than that of typical metals, it is still higher than the conductivity of typical nonmetals, hence their use in semiconductor materials.

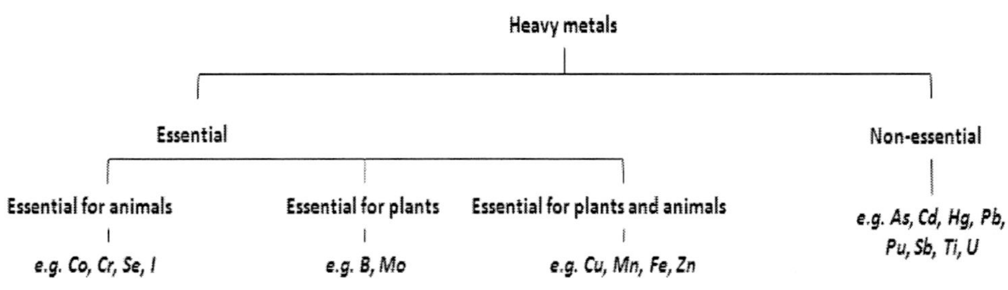

(From Mirsal, 2004)

Figure 1. Diagrammatic representation of the biochemical classification of heavy metals.

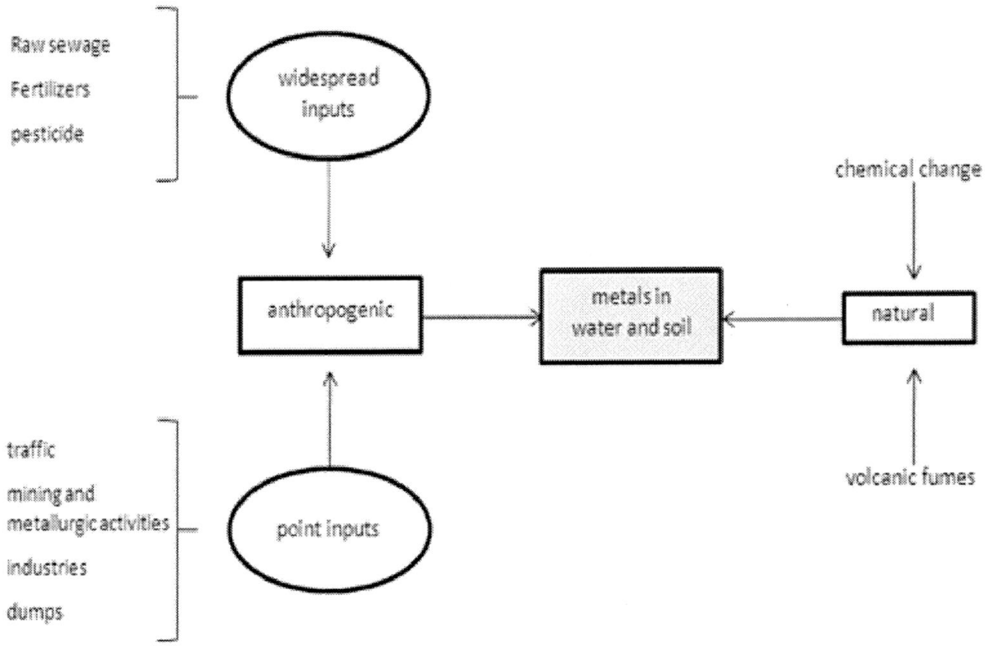

(From Thornton, 1981).

Figure 2. Potential natural (mainly due to alteration of the rocks) and anthropic (human activities) flows of metals in waters and soils.

Their chemical properties are also intermediate, with origins similar to heavy metals. Metalloids may cause enzymatic disorders, skin and cancerous changes and gastrointestinal irritations.

Fruit and Vegetables

The definition of fruits and vegetables generally include the edible parts of plants. Fruit and vegetable have different nutrient contents (Bazzano, 2005).

Fruits and vegetables are a good source of many vitamins and minerals such as folic acid, vitamin C, vitamin E, magnesium and potassium (Lampe, 1999; Nordic Council of Ministers, 2004). They are also a good source of dietary fibres, carotenoids and flavonoids as well as of other bioacative compounds such as plant-sterols (Lampe, 1999; Nordic Council of Ministers, 2004) (Table1a and b).

Antioxidants found in various berries, fruits and vegetables, inactivate reactive oxygen and by that, delay or prevent oxidative damage (Bazzano, 2005).

Stimulation of the immune system, even antibacterial and antiviral activity, modulation of detoxifying enzymes, antioxidant activity, decrease in platelet aggregation, alteration in cholesterol metabolism, modulation of steroid hormone metabolism and blood pressure reduction, have been hypothesized as mechanisms of various intake of fruits and vegetables (Lampe, 1999; Nordic Council of Ministers, 2004).

Controlled trials with micronutrient supplementation have failed to show an effect on chronic disease risk but whole foods, rich in micronutrients e.g. fruits and vegetables show evidence of decreased risk and reduction of chronic diseases (Woodside et al, 2005).

Table 1a. Nutrient content of raw fruits

Nutrient	Nutrient in fruits per 100g raw edible portion
Protein (g)	0.3 (apples) to 2.6 (passionfruit)
Fat (g)	trace (most) to 19.3 (avocado)
Sugars (g)	trace (olive) to 20.9 (bananas)
Dietary fibre (g)	0.1 (watermelon) to 3.8 (kumquat)
β-Carotene (mg)	5.0 (lemon) to 1800 (mango)
Vitamin C (mg)	3.0 (durian) to 230 (guava)
Folate (mg)	3.0 (grapes) to 34 (blackberries)
Vitamin B6 (mg)	0.02 (gooseberries) to 0.29 (banana)
Vitamin E (mg)	0.2 (strawberries) to 2.4 (blackberries)
Potassium (mg)	88.0 (watermelon) to 400 (banana)

(From Holland et al., 1992).

Table 1b. Nutrient content of vegetables

Nutrient	**Nutrient per 100g raw edible portion**
Energy (kJ)	27.0 (lettuce) to 441.0 (taro)
Protein (g)	0.6 (celery) to 4.7 (broccoli)
Fat (g)	0.0 (chinese cabbage) to 0.6 (pumpkin)
Carbohydrate (g)	0.4 (lettuce) to 30.4 (cassava)
Fibre (g)	0.8 (globe artichoke) to 4.0 (broccoli)
β-Carotene (mg)	0.0 (potato) to 10.35 (carrots)
Vitamin C (mg)	4.0 (lettuce) to 110.0 (brussels sprouts)
Calcium (mg)	4.0 (potato) to 82.0 (okra)
Iron (mg)	0.2 (pumpkin) to 3.2 (English spinach)
Flavonoids (mg)	0.0 (spinach) to 35.3 (onion)

(From Hertog et al., 1992).

These compounds can have complementary and overlapping mechanisms of action and a whole variety of mechanisms have been postulated as potential disease-preventive mechanisms (Lampe, 1999). With a higher amount of fruits and vegetables consumption and a more intake variability, the health benefits are increased (The Public Health Institute of Iceland, 2006).

In the early 1990s the long-standing, traditional basic four food groups, consisting of meat, dairy products, grains, and fruits and vegetables, were reworked into a balanced and healthy food guide pyramid. This pyramid has as its base the grain group; on the second level there are the fruit and vegetable groups; on the third level there are the meat and dairy groups; and at the top there is the fats, oils, and sweets' group (Anonymous, 1999). The sources of most vitamins and minerals belong to fruits and vegetables. This pyramid suggests three to five servings of vegetables and two to four servings of fruit should be eaten every day. They also provide fiber, which contains no nutrients but aids in moving food through the digestive system (Figure 3).

The International Agency for Research on Cancer (IARC), which reviews all studies on relationship between fruit and vegetables and cancer, estimates that cancer incidence could decrease by 5-12% with increased fruit and vegetable consumption. A meta-analysis by Riboli and Norat (2003) concluded the following:

Fruit has:

- A protective effect against the risk of cancers of the lung and bladder;
- A protective effect against the risk of cancers of the mouth and pharynx, larynx and oesophagus;
- A protective effect against the risk of cancers of the stomach and colon-rectum;
- No significant protective effect against the risk of breast cancer

Vegetables have:

- A protective effect against the risk of cancers of the oesophagus, breast, lung, stomach, and colorectum;
- No significant protective effect against the risk of cancers of the mouth and pharynx, larynx or bladder.

Thus, there is potential for preventing cancer through diet, particularly through the consumption of fruit and vegetables. The benefits of increased fruit and vegetable consumption in populations also goes beyond specific cancer risks to other chronic diseases.

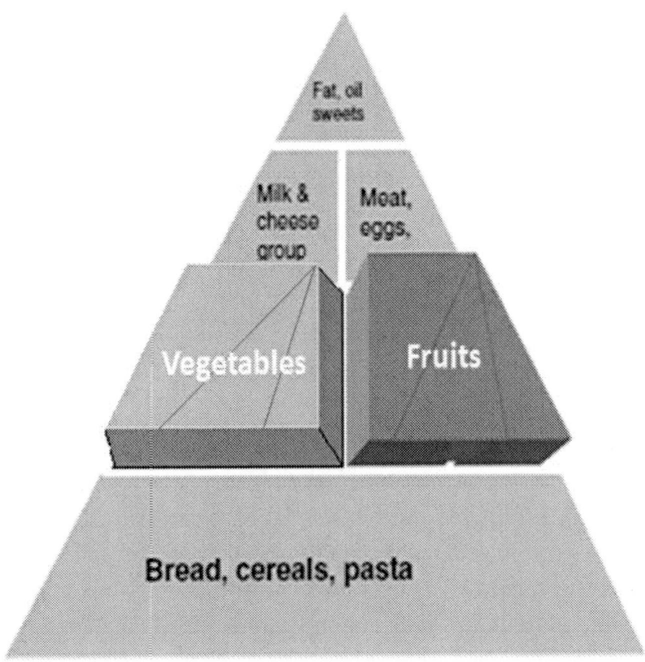

Figure 3. Food guide pyramid.

In this way, fruit and vegetables are important components of (a) the human diet, by contributing, as said before, protein, vitamins, potassium, calcium, iron, and other essential nutrients generally needed in short supply (Thompson and Kelly, 2003), and (b) human health, by contributing to reduce the cancer risk (Riboli and Norat, 2003). So, human beings are encouraged to consume more vegetables and fruits, but, on the other hand, these products can also contain trace metals over a wide range of concentrations, because of the environmental pollution.

This contamination, as already said, is due to irrigation with contaminated water, the addition of fertilizers and metal-based pesticides, industrial emissions, transportation, the harvesting process, storage and/or at the point of sale (Papa et al., 2010 and 2011; Akan et al., 2009; Khairiah et al, 2004; Chojnacka et al, 2005; Stalikas et al. 1997; Mohamed et al. 2003; Sharma et al. 2007). In particular, the plants take them up from soil as well as from air particulate (Papa et al., 2009 and 2011; Khairiah et al, 2004; Chojnacka et al, 2005).

Many studies performed in different areas of the world and in different environmental context showed high content of TMs in vegetables. Vegetables from Addis Ababa market showed that lettuce contained the highest Cd whereas the cabbage accumulate the least (Rahlenbeck et al., 1999). Similar trends in Swiss chard and cabbage from Akaki farm, which were irrigated with industrial effluent (Fisseha and Olsson, 2004). In addition, from the six vegetables grown with industrial and municipal wastes, in Akaki, the highest concentrations of Co were observed in potato followed, in decreased order, by lettuce and Swiss chard. Lettuce accumulated the highest concentrations of Cr and Fe; whereas kale from the same place accumulated highest amounts of Cu, Ni, and Zn. Swiss chard and carrot from Akaki contained the highest Mn and Pb, respectively.

On edible green vegetables grown along the Sinza and Msimbazi rivers of Tanzania revealed that some of the vegetables contained high levels of metals beyond the levels designated by FAO/WHO (Bahemuka and Mubofu. 1999).Alam et al. (2002) showed that some fruit and vegetables in Samta Village in Bangladesh exhibited a potentiality to accumulate high levels of Pb that would pose a health hazard for human consumption. Cui et al. (2004) reported that vegetables and soils from villages located at 1500 m and 500 m near a smelter in Naning, China, were heavily contaminated compared to those coming from a village located at 50 km far from the smelter. In Lagos city, Nigeria, by comparing the metal contamination levels in fruits, vegetables and soils from industrial and residential areas (Yusuf et al. 2003) it has been showed that metal contents were higher in products from industrial areas and the cultivations must be discouraged there. Papa et al. (2011) reported a translocation of Cd, Cr, Ni, Pb, Cu and V contents from soil to the different organs of the tomato plant up to the fruit, despite the root make a barrier for most metals, when the environmental contamination was very high.

Besides, in the UK there are estimated to be over 300,000 allotments (Pless-Mulloli et al., 2004) and urban gardens, often in regions known to have a legacy of environmental pollution depending on former industrial activities, coal burning, motor vehicle emissions, waste incineration and dumping (Pless-Mulloli et al 2004; Alloway, 2004; Prasad and Nazareth, 2000), where fruit and vegetables grow.

In two case studies, in Newcastle and London, despite the soil contaminants considerably exceeding guideline values and the fruit and vegetables produced from the sites revealed heavy metals concentrations, in most samples, exceeding the statutory limit set for commercial food, none of the person, that ate these products, reported any chronic illness.

This is probably due to the fact that some damage may not occur simultaneously (Vandermoere, 2008).

Factors Affecting Heavy Metals Mobility and Bioavailability in Plants

Plant uptake of trace elements is generally the first step of their entry into the agricultural food chain. Plant uptake is dependent on (1) movement of elements from the soil to the plant root, (2) elements crossing the membrane of epidermal cells of the root, (3) transport of elements from the epidermal cells to the xylem, in which a solution of elements is transported from roots to shoots, and (4) possible mobilization, from leaves to storage tissues used as food (seeds, tubers, and fruit), in the phloem transport system. After plant uptake, metals are available to herbivores and man both directly and through the food chain. The limiting step for elemental entry to the food chain is usually from the soil to the root (Chaney and Ryan, 1994).

Plant species and the relative abundance and availability of necessary elements also control metal uptake rates. In fact, different plant species and, within the same species, different cultivars, seems to have different ability to take up heavy metals. For example, in table 2 the content of Cd in different plant species can be seen. (Pettersson, 1994).

Besides, another factor of heavy metal uptake by the plants is the distance with which the metal is transported in the plant. Cadmium, for example, during the route, accumulates in the tissues in different ways (figure4) (Pettersson, 1994). The root of a plant contains higher amounts than the stem, which has higher amount than leaves and fruits. Fruits have generally low amounts of heavy metals if they not have been contaminated through air deposition.

Higher bioavailable of essential nutrients, including P and Ca, can decrease plant uptake of non-essential but chemically similar elements, including As and Cd, respectively. Bioavailability may also be related to the other element availability. For example, Cu toxicity is related to low contents of Zn, Fe, Mo and (or) sulfate (Chaney and Ryan, 1994).

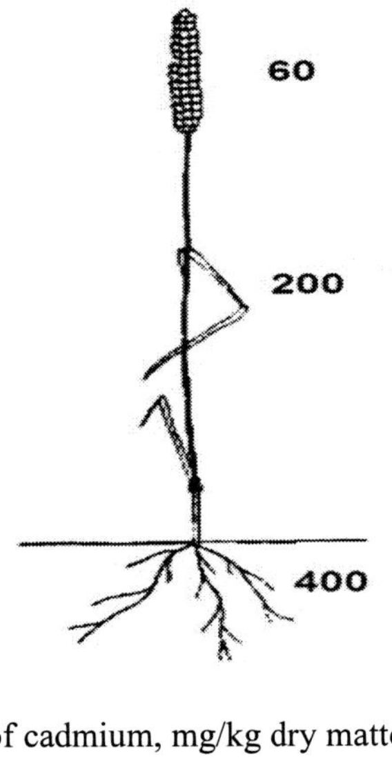

Figure 4. The approximate amounts of cadmium, mg/kg dry matter, in different parts of a wheat plant (Pettersson, 1994).

It is noted that Cd has some similarities with Zn, and is able to mimic Zn that is an essential element. Because of this ability, Cd can replace Zn in the uptake and metabolic functions of the plant (Moolenaar, 1998). In particular, Cd interacts with the metabolism of four metals essential to nutrition: Zn, Fe, Ca and Cu. Table 3 shows the major Cd interactions with these micronutrients and other dietary components.

Table 2. Normal Occurring Cadmium-Content, Mg/Kg Dry Matter, in Different Plant Species (Pettersson, 1994)

Plant species	Average cadmium content µg/KgDM
Salad	8
Carrot	22
Potato	17
Apple	<1
Strawberries	8
Sunflower seeds, peeled	380
Autumn wheat	60
Oat	40
Barley	20
Rye	15

(Pettersson, 1994).

Table 3. Major Cadmium Interactions with Micronutrients and Other Dietary Components(from Marjorie and Peraza, 1998)

Metal		Toxicity
Cd		Anemia; osteoporosis; proximal tubular disfunction leading to hypertension, coronary artery disease, and chronic pulmonary diseases.
Metal-nutrient	**Interaction and mechanism**	**Effect of nutrient on metal toxicity**
Cd-zinc	Competes for GI absorption; Cd interferes with zinc metabolism.	Reverses Cd toxicity (i.e., decreases growth, increases lesions and testicular necrosis).
Cd-iron	Cd decreases iron absorption and metabolism (Cd possibly binds with ferritin and transferrin).	Supplementation corrects anemia: increases hematocrit and increases hemoglobin levels.
Cd-calcium	Cd decreases intestinal calcium transport; increases Cd deposits in bone tissue in a calcium-deficient state.	Sufficiency protects against bone deformities, osteomalacia, and osteoporosis (Itai-ltai disease).
Cd-copper	Cd interferes with copper metabolism, possibly by decreasing copper absorption.	Corrects Cd-induced decreased plasma ceruloplasmin concentrations.
Cd-protein	Low-protein diet results in increase Cd uptake.	Sufficiency prevents Cd-induced decreased growth, decreases MT synthesis, and increases bone deformities.
Cd-selenium	Selenium shifts Cd binding to higher molecular weight proteins	MT can now bind essential nutrients.

GI, gastrointestinal.
(From Marjorie and Peraza, 1998).

The bioavailability of elements to plants is also controlled by many factors associated with soil and climatic conditions, plant genotype and agronomic management, including total concentration and speciation (physical-chemical forms) of metals, mineralogy, pH, redox potential, temperature, total organic content, and suspended particulate content, the type of plant root system and the response of plants to elements in relation to seasonal cycles (Kabata-Pendias and Pendias, 1992).

Many of these factors vary seasonally and temporally, and most of them are interrelated. Consequently, the change of one factor may affect several others. In addition, other poorly understood biological factors seem to strongly influence bioaccumulation of metals and severely inhibit prediction of metal bioavailability (Luoma, 1989).

Toxic Effects of Trace Metals

Once into the food chain and absorbed by the human body, the trace metals may cause toxicity.

Studies confirm that trace metals can directly influence behaviour by impairing mental and neurological function, influencing neurotransmitter production and utilization and altering numerous metabolic body processes. Systems in which toxic metal elements can induce impairment and dysfunction include the blood, cardiovascular, detoxification pathways, endocrine, energy production pathways, enzymatic, gastrointestinal, immune, nervous, reproductive and urinary (Kellas and Dworkin, 1996). When ingested and inhaled in excessive amounts, heavy metals can affect the liver, brain and lungs, although each metal causes its own characteristic symptoms (Friedman, 1994).

Children appear to be more sensitive to heavy metal exposure than adults, and are consequently the focus of concern. The toxic nonessential metals such as Cd and Pb are characterized as having no demonstrated biological requirements in man, and exposure is associated with recognizable toxicity. Also, severity of toxicity increases with increases in dosage (Goyer, 1997). Today, emphasis is being placed on the carcinogenic effects of metals. Pb and Cd are all proven or suspected causes of certain cancers associated with industrial processes.

A metal is regarded as toxic if it injures the growth or metabolism of cells when it is present above a given concentration. Casdorph and Walker (1995) states that almost all metals are toxic at high concentrations, and some are severe poisons even at very low concentrations. Each metal has an optimum range of concentration, in excess of which the element is toxic. The heavy metals listed in the table 4 are amongst the most prevalent. The health effects of exposure to these metals are also identified.

The toxicity of a metal depends on its route of administration and the chemical compound with which it is bound. The combining of a metal with an organic compound may either increase or decrease its toxic effects on cells.

On the other hand, the combination of the metal with S to form a sulphide results in a less toxic compound than the corresponding hydroxide or oxide, because the sulphide is less soluble in body fluids than the oxide. Toxicity generally results when an excessive concentration is presented to an organism a) over a prolonged period of time; b) when the metal is presented in an unusual biochemical form; c) by way of an unusual route of intake.

Table 4. Examples of health effects of exposure to heavy metals

Heavy Metal	Result of Exposure
Chromium	Cancer, damage to respiratory organs
Cadmium	Anemia, emphysema, fatigue, hair loss, heart disease, hypertension, kidney and liver damage, lung cancer
Lead	Brain damage, anorexia, anemia, constipation, convulsions, miscarriage
Nickel	Cancer, contact dermatitis, diarrhea, headaches, skin rashes, nausea

In order to avoid undesirable health as resulted of "excessive" intake of toxicants (including toxic metals), international and national scientific organisms such as FAO/WHO, FDA, European Union, etc. have used the safety factor approach for establishing acceptable or tolerable intakes of substances that exhibit threshold toxicity. The acceptable daily intake (ADI) or tolerable daily intake (TDI) or provisional tolerable weekly intakes (PTWI) are used to describe "safe" levels of intake for several toxicants including toxic metals (Olivera da Silva et al., 2005).

A more detailed discussion of the health effects of metals is presented below.

Copper

Copper is an essential element to all living organisms, and because of that both deficiency and excess have consequence for the integrity of biochemical functions. The main biological role of copper is as an ingredient, normally in the prosthetic group, of oxidising enzymes which are important in oxidation-reduction processes (Moolenaar, 1998). Copper is highly toxic when in excess (de Bie et al., 2005). Even small amounts of copper, such as the dose contained in drinking water, could lead to toxic effects. These small amounts of copper can cause nausea (Hotz et al.,2003). When copper is consumed in excess it can have a variety of symptoms such as mental diseases, learning disabilities, behaviour problems, tiredness, muscle and joint pain, irritability, depression, nervousness and fatigue. It has also been found that premature wrinkling of the skin is related to excess consumption of copper (Campbell, 2001).

Copper deficiency on the other hand may contribute to week blood vessels (aneurism), an enlarged heart, elevated cholesterol, gout, hyperactivity, emotional disturbance and depressed dopamine levels. A shortage of dopamine may result in defects related to multiple sclerosis (MS) and Parkinson's disease (Campbell, 2001). Copper deficiency can also cause an iron-deficiency-type of anaemia (Aaseth and Norseth, 1986).

Joint FAO/WHO Expert Committee on Food Additives (JECFA) (1982) has established a Provisionally Maximum Tolerable Daily Intake (PMTDI) of 0.5 mg/kg/b.w./day.

Chromium

Chromium is an essential nutrient to the human body as it is required to promote the action of insulin for the utilization of sugars, proteins and fats. Chromium is a naturally occurring heavy metal that is found in the environment as trivalent, Cr (III) and hexavalent, Cr (VI) forms, where Cr (VI) is responsible for most of the toxic actions (Shrivastava et al., 2002). A high dose of chromium and long term exposure can give rise to various, cytotoxic and genotoxic reactions that affect the immune system and may result in immuno-stimulation or immuno-suppression. Occupational exposure to chromium is mainly through inhalation and non-occupational exposure is mainly through consumption. Chromium is very toxic when inhaled and causes lung cancer, nasal irritation, nasal ulcer, hypersensitivity reactions and contact dermatitis. When chromium is ingested its mainly Cr (VI) that is toxic, although not as toxic as through inhalation. When Cr (VI) is digested it is reduced to Cr (III) before entering the blood stream. The reduction of Cr (VI) to Cr (III) results in the formation of reactive intermediates that together with oxidative stress and oxidative tissue damage and a number of other molecular events contribute to the cytotoxicity, genotoxicity and carcinogenicity of Cr (VI)-containing compounds (Shrivastava et al., 2002). Chromium salts, chromium polynicotinate, chromium chloride and chromium picolinate (CrP) are used as micronutrients and nutritional supplements and have been demonstrated to possess a number of health benefits to humans and animals. CrP for example have been used as a nutritional supplement to control blood sugar in diabetes patients and it may also reduce cholesterol and blood pressure levels (Shrivastava et al., 2002). The general non-occupational exposure of the population to chromium is from food such as vegetables and meat. Other ways of non-occupational exposure is through urban air, hip or knee prostheses and cigarettes (Shrivastava et al., 2002).

Joint FAO/WHO Expert Committee on Food Additives (JECFA) has not evaluated chromium. The National Science Academic of Washington in 1993 indicated a PTWI of 0.35-1.40 mg/Kg/b.w./week.

Zinc

Zinc is an essential nutrient for the human body and has an importance for health (Hotz et al., 2003). Zinc acts as a catalytic or structural component in many enzymes that are involved in energy metabolism and in transcription and translation of RNA (Moolenaar, 1998). Deficiency symptoms in humans and animals are failure to growth, loss of apetite, acne, slow healing, strong body and breath odour, and a decrease in taste perception (Campbell, 2001) Zinc also has a prominent role in determining the outcome of pregnancies and supporting neurobehavioral development (Hotz et al., 2003). Zinc deficiency is also closely correlated to optimum blood circulation and cardiovascular disease (Campbell, 2001). There is a lot of information available on zinc deficiency; however there is relatively little information on toxic effects caused by zinc excess (Hotz et al., 2003). Joint FAO/WHO Expert Committee on Food Additives (JECFA) (1982) has established a PMTDI of 1mg/Kg/b.w./day

Nickel

Since nickel is not proven to be an essential element in humans it is not clear how nickel compounds are metabolized in the body. It is however known that over exposure to nickel compounds may have an adverse effect on human health. The most common and well known reaction of nickel on humans is contact-allergy. Nickel can also be accumulated in the body through chronic exposure, which can cause lung fibrosis and cardiovascular- and kidney diseases. However the most serious concern is nickel's carcinogenic effect and its ability to induce tumours at nearly any site of administration (Denkhaus and Salnikow, 2002). The primarily exposure of the population to nickel is via inhalation and ingestion. Some vegetables such as spinach contain high amounts of nickel. Other dietary exposure to nickel is through cocoa and nuts which also contain high amounts. It is however very difficult to maintain a nickel-deficient diet since it is present in all types of food (Denkhaus and Salnikow, 2002). In every case, WHO (1997) has given a TDI of 0.005 mg/kg/b.w./day.

Mercury

There are three forms of mercury and among these the most toxic one is the organic form, methyl mercury. Methyl mercury is microbiologically transformed from inorganic mercury when it reaches aquatic environments, in water bodies or in soils (Zahir et al., 2005). Inorganic- and organic mercury is toxic to the human body in different ways, effecting different organs in different ways. Inorganic mercury can cause neurological and psychological symptoms, such as tremor, changes in personality, restlessness, anxiety, sleep disturbance and depression. These symptoms are however reversible after ending of exposure to inorganic mercury. Inorganic mercury is also an allergen, which may cause contact eczema. The kidneys are the organs that accumulate the highest levels of mercury compared to brain and liver. This can cause kidney damage which is reversible after the exposure has stopped.

Organic mercury, methyl mercury, toxicity is not reversible as it is with inorganic mercury. Organic mercury affects the nervous system and the main symptoms of methyl mercury poisoning relate to damage of the nervous system. The earliest symptoms of poisoning are parestesias and numbness in the hands and feet. Later symptoms are coordination difficulties and concentric constriction of the visual field (Järup, 2003). Other symptoms are memory loss, shortfall in attention and Alzheimer's disease like dementia (Zahir et al., 2005). Hock et al. (1998) made a study on if environmental factors may influence the risk of getting Alzheimer's disease. They found that Alzheimer's disease patients had a two-fold higher blood-mercury level than the control group and that early onset Alzheimer's disease patients, blood-mercury levels where three-fold higher than the control group. Exposure to the foetus of humans to mercury can also cause late development of speech, late walking, memory shortfall in attention and Autism (Zahir et al., 2005) The general population is primarily exposed to mercury via food, where fish is the major source of methyl mercury exposure (Järup, 2003). However some vegetables have been found to be able to accumulate mercury at significant levels, such as *Brassica oleracea* and *Amaranthus oleraceous* (Zahir et al., 2005). Joint FAO/WHO Expert Committee on Food Additives (JECFA, 1978 and 1988) has established a Provisionally Tolerable Weekly Intake (PTWI) at

0.005 mg/kg/b.w./week for mercury, however, with a maximum at 0.0033 mg/kg/b.w./week for methyl mercury. It was, however, stated that this PTWI might not adequately protect pregnant women, i.e. the foetus.

Arsenic

Significant exposure to arsenic occurs through both anthropogenic and natural sources. Occupational and community exposures to arsenic from the activities of humans occur through the smelting industry, the use of gallium arsenide in the microelectronics industry, and the use of arsenic in common products such as wood preservatives, pesticides, herbicides, fungicides, and paints. Widespread dispersion of arsenic is a byproduct of the combustion of fossil fuels in which arsenic is a common contaminant.

The toxicity of an arsenic-containing compound depends on its valence state (zero-valent, trivalent, or pentavalent), its form (inorganic or organic), and factors that modify its absorption and elimination. Inorganic arsenic is generally more toxic than arsenic, and trivalent arsenite is more toxic than pentavalent and zero-valent arsenic. These nuances are important. For example, testing biological samples for arsenic in an individual with suspected toxicity must be done more than 48 hours after the individual abstains from eating seafood; otherwise, the test may be confounded by the presence of arsenobentaine, a relatively harmless form of arsenic that is contained in fish at high levels of concentration.

Once absorbed into the body, arsenic undergoes some accumulation in soft tissue organs such as the liver, spleen, kidneys, and lungs, but the major long-term storage site for arsenic is keratin-rich tissues, such as skin, hair, and nails---making the measurement of arsenic in these biological specimens useful for estimating total arsenic burden and long-term exposure under certain circumstances.

Acute arsenic poisoning is infamous for its lethality, which stems from arsenic's destruction of the integrity of blood vessels and gastrointestinal tissue and its effect on the heart and brain. Chronic exposure to lower levels of arsenic results in somewhat unusual patterns of skin hyperpigmentation, peripheral nerve damage manifesting as numbness, tingling, and weakness in the hands and feet, diabetes, and blood vessel damage resulting in a gangrenous condition affecting the extremities (Col et al., 1999). Chronic arsenic exposure also causes a markedly elevated risk for developing a number of cancers, most notably skin cancer, cancers of the liver (angiosarcoma), lung, bladder, and possibly the kidney and colon. The dose necessary to increase the risk for cancer has recently become the focus of particularly intense scrutiny in the U.S. because of proposed efforts to lower standards governing general populations exposures to arsenic. Among environmental scientists studying this problem, the most common view is that the current standard for the allowable amount of arsenic in U.S. drinking water—50 mg/liter—is probably not adequate to sufficiently safeguard the general population from arsenic's cancer risk (Morales et al., 2000).

Lead

Lead is toxic because it mimics many aspects of the metabolic behaviour of calcium and inhibits many enzyme systems. In man, one of the chief concerns in lead toxicity is its effect in causing brain damage particularly to children (Tsuchiya, 1986). Children are very susceptible to lead exposure (Järup, 2003). Some chronic effects are related to the haematopoietic system (anaemia), the nervous system (encephalopathy), gastrointestinal tract (stomach complaints etc) and the kidneys (renal tubular dysfunction) (Tsuchiya, 1986). The general exposure to lead is from air and food in more or less equal proportions (Järup, 2003). There are few reports on lead induced toxic effects on plant growth in natural ecosystems that have been severely polluted with lead (Koeppe, 1981). This is because lead is consequently unavailable to plants (Brady and Weil, 2002). The more significant role of plants with Pb in the ecosystem resides in the food chain, where deposition of lead on the leaf surface may be ingested by herbivores (Koeppe, 1981).

Joint FAO/WHO Expert Committee on Food Additives (JECFA, 1993) has established a PTWI at 0.025 mg/kg/b.w./week. Estimated daily dietary intake for adults range from 0.015-0.1 mg, depending on the composition of the diet and where the consumer lives (Codex, 1995).

Cadmium

Cadmium is toxic because it has some similarities with zinc that is an essential element. Because of this, cadmium have the ability to mimic zinc in some metabolic functions, but unlike zinc there has been no evidence that cadmium is essential to any metabolic function. Instead cadmium has the ability to disturb enzyme activity in the body (Moolenaar, 1998). In a study done by Järup and Carlsson (1999) it showed that exposure to cadmium was connected to damage on the kidneys as well as kidney stone (renal calculus). Cadmium is probably also carcinogenic to humans. At long-term exposure to high cadmium levels, cadmium may cause skeletal damage (osteoporosis and osteomalacia) (Järup, 2003).

Many investigations done in Sweden the resent years have suggested that the population exposure to cadmium is increasing. This is of course alarming since the current exposure is near the level for minor changes in the kidneys function begin to appear (tubular protein urea) (Bäcklin and Thuvander, 1999). Adequate levels of essential elements especially zinc and vitamin C reduces abortion and excretion of cadmium (Campbell, 2001). Vegetables, root crops, cereals and offal are the foods that the most contribute to the accumulation of cadmium in the human body. For an average consumer one third of the intake of cadmium is retrieved through cereal products and another third through vegetables and root crops (Pettersson, 1994b). Scientific Committee on Food of the European Union (SCF, 1997) endorsed the JECFA PTWI of 0.007 mg/kg/b.w./week. The committee concluded that for a significant part of the population, exposure to cadmium from dietary sources alone is at a level close to the PTWI.

Manganese

Biological Mn is considered to be an essential metal important to mitochondrial oxidative processes for all living mammals, but may also be toxic at high concentrations (Rollin et al., 2005). Excessive intake of Mn, either through inhalation or ingestion, may result in pathology, particularly to the central nervous system. Excessive exposure via the inhalation has been shown to cause effects on the lungs and accumulate in the brain, causing irreversible brain disease (Rollin et al., 2005). Joint FAO/WHO Expert Committee on Food Additives has not evaluated manganese. World Health Organization (WHO, 1993) recommends a daily intake of 2-3 mg/Kg/b.w./day.

Iron

Iron is physiologically essential for life, but biochemically dangerous. Toxicity attributable to excess Fe can occur chronically because of excessive accumulation in the body from contaminated food sources. Iron toxicity can have a direct effect on the gastrointestinal tract and can cause death due to widespread cellular dysfunction (Hardman and Limbird, 1995). Joint FAO/WHO Expert Committee on Food Additives (JECFA, 1983) has established a PMTDI at 0.8 mg/kg/b.w./day. The value applies to iron from all sources except for iron oxides used as coloring agents, supplemental iron taken during pregnancy and lactation and supplemental iron for specific clinical requirements. The value is 8 times lower than the acute toxic dose.

Conclusion

Fruit and vegetables contamination by metals are now beginning to be a real risk to human health, despite the rigorous regulations at Community level requiring many controls and heavy penalties for non-compliance of the directives. It's, therefore, of vital importance groped to reduce emissions of these elements in the environment developing more environmentally technologies (eg, the replacement of leaded premium gasoline with a green one, have significantly decreased the release of this metal), as contamination the food chain is mainly due to the diffusion of metals in the environment in quantities that far exceed those due to natural phenomena such as erosion or eruptions, causing accumulation in the environment. All these metals, in fact, have toxic potential, but the detrimental impact becomes apparent only after decades of exposure. It is, therefore, suggested that regular monitoring of heavy metals in plant tissues is essential in order to prevent excessive build-up of these metals in the human food chain.

References

Aaseth, J., Norseth, T., 1986. Cupper. Ingåri: Friberg, L. Nordberg, G. F. and Vouk, V. B. (Red); Handbook on the toxicology of metals. Vol. II: Specific metals. *Elsevier Science Publishers B.V.* Amsterdam, pp 233-253.

Akan J.C., Abdulrahman F. I., Ogugbuaja, V. O., Ayodele J. T., 2009. Heavy Metals and Anion Levels in Some Samples of Vegetable Grown Within the Vicinity of Challawa Industrial Area, Kano State, Nigeria. *American Journal of Applied Sciences.* 6 (3): 534-542.

Alam, M.G.M., Snow E.T., Tanaka A., 2002. Arsenic and heavy metal contamination of vegetables grown in Samta village, Bangladesh. *Sci. Total Environ.* 308: 83-96.

Alloway BJ, 2004. Contamination of soils in domestic gardens and allotments: a brief overview. *Land Contamination and Reclamation.* 12:179-187.

Alloway, B.J., 1995. Heavy Metals in Soils. London: *Blackie Academic and Professional. p. 141.*

Anonymous, 1999. Position of the American Dietetic Association: functional foods. *J. Am. Diet Assoc.* 99, 1278–85.

Bäcklin, B-M., Thuvander, A., 1999. Miljö och folkhälsa (Rapport 4904). Statens naturvårdsverk. Heavy metals in edible green vegetables grown along the sites of the Sinza and Msimbazi rivers in Darer Salaam, Tanzania. *Food Chem.* 66: 63-66.

Bahemuka, T. E., Mubofu, E.B., 1999. Heavy metals in edible green vegetables grown along the sites of the Sinza and Msimbazi rivers in Dares Salaam, Tanzania. *Food Chemistry.* Vol. 66, pp. 63-66.

Bazzano, L. A., 2005. Dietary intake of fruit and vegetables and risk of diabetes mellitus and cardiovascular diseases [electronic resources]. Kobe, Japan.

Bower, C., 1994. Cadmium in vegetables. Monthly Newsletter No. 110 (june), NSW Agriculture, Orange, Australia Campbell, J. D., 2001. Lifestyle, minerals and health. *Medical Hypotheses.* vol. 57, no.5:pp521-531.

Casdorph, H., Walker, M., 1995. Toxic Metal Syndrome. New York: *Avery Publishing.* p. 95.

Chaney, R. L., Ryan, J. A., 1994. Risk based standards for arsenic, lead and cadmium in urban soils. Summary of methods developed to estimate standards for Cd, Pb and As in urban soils. Frankfurt am Main: DECHEMA.

Charya N.S., Kamalaa C.T., Rajb D.S.S.,2008. Assessing risk of heavy metals from consuming food grown on sewage irrigated soils and food chain transfer. *Ecotoxicology and Environmental Safety.* 69: 513–524.

Chojnacka K, Chojnacki A, Gorecka H, Gorecki H. Bioavailability of heavy metals from polluted soils to plants. *Sci. Total Environ.* 2005; 337:175-82.

Codex Alimentarius Commission, 1995. Doc. no. CX/FAC 96/17. Joint FAO/WHO food standards programme. Codex general standard for contaminants and toxins in foods.

Col, M, Col, C, Soran, A, Sayli, BS, Ozturk, S., 1999. Arsenic-related Bowen's disease, palmar keratosis, and skin cancer. *Environ. Health Perspect.*107:687-9.

Cui, Y. J., Y.G. Zhu, R.H. Zhai, D.Y. Chen, Y.Z. Huang, Y. Qiu and J.Z. Liang, 2004. Transfer of metals from soil to vegetables in an area near a smelter in Naning, China. *Environ. Int.* 30: 785-791.

de Bie, P., van de Sluis, B., Klomp, L., Wijmenga, C., 2005. The many faces of the copper metabolism protein MURRI/COMMDI. *Journal of Heredity.* vol 96, no 7: pp 803-811.

Denkhaus, E., Salnikow, K., 2002. Nickel essentiality, toxicity and carcinogenicity. *Critical Reviews in Oncology/Hematology.* vol 42, pp 35-56.

Ekholm P., Reinivuob H., et al., 2007. Changes in the mineral and trace element contents of cereals, fruits and vegetables in Finland. *Journal of Food Composition and Analysis.* 20: 487–495.

Fisseha I, Olsson M., 2004. Land Degradation in Addis in Ababa due to Industrial and Urban Development. *Ethiopian Journal of Developmental Research.* 26:77-100.

Friedman, K., 1994. Reporting on the Environment: A handbook for Journalists. p. 8.

Goyer, RA, 1997. Toxic and Essential metal interactions. *Annual Review of nutrition.* Vol. 17, pp. 37-50.

Hardman, J.G. and Limbird, L.E, 1995. The Pharmacological Basis ofToxicitv. McGraw-HiII, New York. p. 1324.

Hertog, M.G.L., Hollman, P.C.H., Katan, M.B., 1992. Intake of potentially anticarcinogenic flavonoids and their determinants in adults in the Netherlands. *J. Agric. Fd Chem.* 41, 1242–6.

Hock, C., Drasch, G., Golombowski, S., Müller-Spahn, F., Willerhausen-Zönnchen, B., Schwarz, P., Hock, U., Growdon, J. H., Nitsch, R. M., 1998. Increased blood mercury levels in patients with Alzheimer's disease. *Journal of Neural Transmission.* (Vol 105). pp 59-68.

Holland, B., Unwin, I.D., Buss, D.H., 1992. Fruit and nuts. First Supplement to McCance and Widdowson's, the composition of foods (5th edition). *Royal Society of Chemistry.* MAFF, London.

Hotz, C., Lowe, N. M., Araya, M. , Brown, K. H., 2003. Assessment of the trace element status of individuals and popultions: The example of zinc and copper. *The American society of nutritional science.* vol 133, no 5: pp 1563-1560.

Huang M., Zhoub S., Sunc B., Zhaoc Q., 2008. Heavy metals in wheat grain: Assessment of potential health risk for inhabitants in Kunshan, China. *Science of the total environment.* 405: 54-61.

Järup, L., Carlsson, D., 1999. Benskörhet och njurskador efter miljöexponering för kadmium I arbetet och omgivningen. Ingår i Bäcklin, B-M. and Thuvander, A. (Red.) Miljö och folkhälsa (Rapport 4904). Statens Naturvårdsvek p 76-83.

Järup, L., 2003. Hazards of metal contamination. Ingår i British medical bulletin. *The British Council.* vol. 68: pp 167-182.

JECFA, 1978. Evaluation of certain food additives and contaminants. Twenty-second report of the Joint FAO/WHO Expert Committee on Food Additives. World Health Organization, *Technical Report Series.* 631.

JECFA, 1982. Evaluation of certain food additives and contaminants. Twenty-sixth report of the Joint FAO/WHO Expert Committee on Food Additives. World Health Organization, Technical Report Series 683.

JECFA, 1983. Evaluation of certain food additives and contaminants. Twenty-seventh report of the Joint FAO/WHO Expert Committee on Food Additives. World Health Organization, *Technical Report Series.* 696.

JECFA, 1988. Evaluation of certain food additives and contaminants. Thirty-third report of the Joint FAO/WHO Expert Committee on Food Additives. World Health Organization, *Technical Report Series*. 776.

JECFA, 1993. Evaluation of certain food additives and contaminants. Forty-first report of the Joint FAO/WHO Expert Committee on Food Additives. World Health Organization, *Technical Report Series*. 837.

Kabata–Pendias, A., Pendias H., 1992. Trace Elements in Soil and Plants. 2nd ed. Boca Raton. Florida, USA:CRC Press.

Kellas, B., Dworkin, A., 1996. Surviving the Toxic Crisis. Olivenhain: *Professional Reference Publishing.* pp. 187-230.

Khairiah J., Zalifah MK., Yin YH., Aminha A., 2004. The uptake of heavy metals by fruit type vegetable grown in selected agricultural areas. *Pak. J. Biol. Sci.* 7: 1438-42.

Koeppe, D. E., 1981. Lead: Understanding the minimla toxicity of lead in plants. Ingår i: Lepp, N.W. (Red.); Effect o heavy metal pollutionon plants, Vol. 1, Effect of trace metals on plant function. Pollution monitoring series. Applied science Publishers, London, p 55-76.

Lampe, J. W., 1999. Health effects of vegetables and fruit: assessing mechanisms of action in human experimental studies. *Am. J. Clin. Nutr.* 70(3 Suppl), 475S-490S.

Luoma, S.N. 1989. Can we determine the biological availability of sediment-bound trace elements? *Hydrobiologia.* v. 176/177, p. 379-396.

Marjorie A. Peraza et al., 1998. Marjorie A. Peraza, Felix Ayala-Fierro, David S. Barber, Elizabeth Casarez, and Leonard T. Rael Effects of Micronutrients on Metal Toxicity. *Environmental Health Perspectives.* Vol 106, Supplement I.

Mirsal, I. A., 2004. Soil Pollution: Origin, Monitoring and Remediation. Springer-Verlag Berlin Heidelberg.

Moolenaar, S. W., 1998. Sustainable Management of Heavy Metals in Agro-ecosystems. Landbouwuniversiteit Wageningen.

Morales KH, Ryan L, Kuo TL, Wu MM, Chen CJ, 2000. Risk of internal cancers from arsenic in drinking water. *Environ. Health Perspect.* 108:655-61.

Muchuweti M., Birkett J.W., Chinyanga E., Zvauya R., Scrimshaw M.D., Lester J.N., 2006. Heavy metal content of vegetables irrigated with mixtures of wastewater and sewage sludge in Zimbabwe:Implications for human health. *Agriculture, Ecosystems and Environment.* 112: 41–48.

Nordic Council of Ministers (Ed.)., 2004. Nordic Nutrition Recommendations 2004: Integrating nutrition and physical activity (4 edition ed.). Copenhagen: Nordic Conucil of Ministers.

Olivera da Silva A. L., Barrocas P. R.G., do Counto S.J., Costa Moreira J., 2005. Dietary intake and health effects of selected toxic elements. *Braz. J. Plant Physiol.* 17(1), 79-93.

Papa, S., Bartoli, G., Di Martino, D., Fioretto, A., 2010. Occurrence of platinum in the leaves of holm-oak (Quercus ilex L.) from different sites (streets and squares) in the city of Caserta (Italy). *Fresenius Environmental Bullettin.* vol. 19 – No 9b, p. 2109-2115, ISSN: 1018-4619.

Papa, S., Bartoli, G., Fioretto, A., 2011. Soil-Plant transfer of trace metals to vegetable of tomato plants. In: MESAEP, XVI International symposium on environmental pollution and its impact on life in the Mediterranean region. *Ioannina.* (Greece), September 24-27, 2011.

Papa S., Cerullo L., Di Monaco A., Bartoli G., Fioretto A., 2009. Trace elements in fruit and vegetable products. In: Dassenakis M., Pattee E., Vianello G., Vittori Antisari L.; Environmental Quality p. 199-204

Pettersson, O., 1994. Tungmetaller Ioding ochmiljö. Ingår iserien: Aktuellt från lantbruksuniversitetet, SLU Info/Redaktionen.

Pless-Mulloli T, Air V, Vizard C, Singleton I, Rimmer D, Hartley P, 2004. The legacy of historic land-use in allotment gardens in industrial urban settings: Walker Road allotment in Newcastle upon Tyne, UK. *Land Contamination and Reclamation.* 12:239-251.

Prasad LR, Nazareth B, 2000. Contamination of allotment soil with lead: managing potential risks to health. *Journal of Public Health Medicine.* 22:525-530.

Radwan, M. A., A.K. Salama, 2006. Market basket survey for some heavy metals in Egyptian fruits and vegetables. *Food. Chem. Toxicol.* 44: 1273-1278.

Rahlenbeck, S.I., Burberg, A., Zimmermann, R.D., 1999. Lead and Cadmium in Ethiopian Vegetables. *Bulletin of Environmental Contamination and Toxicology.* Vol. 62, pp. 30-33.

Riboli A., Norat B., 2003. Publication anticipated end of 2003 in the *American Journal of Clinical Nutrition.*

Rollin, H., Mathee, A, Levin, J., Theodorou, P., Wewers, F., 2005. Blood manganese concentrations among firstgrade schoolchildren in two South African cities. *Environmental Research.* Vo1.97, pp. 93-99.

SCF, 1993. Reports of the Scientific Committee for Food. Thirty-first series. Nutrient and energy intakes for the European community.

SCF, 1997. Opinion on cadmium. Reports of the Scientific Committee for Food, Thirty-sixth series.

Shrivastava, R., Upreti, R. K., Seth, P. K., Chaturvedi, U. C., 2002. Effects of chromium on the immune system. FEMS *Immunology and Medical Microbiology.* vol 34, pp 1-7.

The Public Health Institute of Iceland (Ed.), 2006. [Recommendations on diet and nutrients for adults and children from 2 years of age] [booklet in Icelandic]. Reykjavik: Lydheilsustod.

Thompson, HC., Kelly, WC., 2003. Vegetable Crops. 5th Edn. New Delhi: McGraw Hill Publishing Company Ltd; 199067.

Thornton I. Geochemical aspects of the distribution and forms of heavy metals in soils. In:Lepp NW. 1981, Effect of heavy metal pollution on plants: metals in the environment, vol. II. *Applied Sci. Publ., London and New Jersey.* p. 1–34.

Tsuchiya K., 1986. Lead. Ingåri: Friberg, L. Nordberg, G. F. and Vouk, V. B. (Red); Handbook on the toxicology of metals. Vol. II: *Specific metals. Elsevier Science Publishers B.V.* Amsterdam, pp 298-253.

Vandermoere F., 2008. Psychosocial health of residents exposed to soil pollution in a Flemish neighbourhood. *Social Science and Medicine.* 66:1646-1657.

Whedepohl K.H. Ed, 1969. Handbook oh Geochemistry. Springer-Verlag, Berlin.

WHO, 1997. Guideline for drinking water quality, second ed., volume 1, Recommendations.

WHO, 1993. Evaluation of certain food additives and contaminants. *Technical report.* Series 751, pp. 35 - 53.

WHO, 2003. Diet, Nutrition and the Prevention of Chronic Diseases. Report of a Joint WHO/FAO Exper Consutation. Geneva: World Health Organisation.

Woodside, J. V., McCall, D., McGartland, C., Young, I. S., 2005. Micronutrients: dietary intake v. supplement use. *Proc. Nutr. Soc.* 64(4), 543-553.

Xiu-Zhen H., Dong-Mei Z., De-Qian H., Long C., Hai-Lin Z., Hui W., 2009. Heavy Metal Transfer from Soil to Vegetable in Southern Jiangsu Province, China. *Pedosphere.* 19(3): 305–311.

Yusuf, A. A., Arowolo T.A., Bamgbose O., 2003. Cadmium, copper, and nickel levels in vegetables fromindustrial and residential areas of LagosCity,Nigeria. *Food.Chem. Toxicol.* 41:375-378.

Zahir, F., Rizwi, S. J., Haq, S. K., Khan, R. H., 2005. Low dose mercury toxicity and human health. *Environmental Toxicology and Pharmacology.* vol 20, pp 351-360.

In: Trace Elements ISBN: 978-1-62081-376-8
Editors: Diego A. De Leon and Paloma Raquel Aragon

Chapter 4

IMPLICATIONS OF EXCESSIVE LIVESTOCK MINERAL SUPPLEMENTATION ON ENVIRONMENTAL POLLUTION AND HUMAN HEALTH

Marta López-Alonso[1] and Marta Miranda[2]
[1]Universidade de Santiago de Compostela,
Departamento de Patoloxía Animal,
Facultade de Veterinaria, Lugo, Spain
[2]Universidade de Santiago de Compostela, Departamento de Ciencias Clínicas
Veterinarias, Facultade de Veterinaria, Lugo, Spain

ABSTRACT

Trace elements are essential to maintain physiological functions and it is well demonstrated that certain mineral supplementation in livestock can improve productivity. Mineral supplements are routinely incorporated into concentrate feed, even when not necessary, at doses largely (up to 10-fold) exceeding the physiological requirements; this is possible because there are fairly wide "safety margins" to avoid mineral toxicity, even though toxic effects on animals can appear when given at supra-optimal concentrations. A significant proportion of minerals that are given to livestock are excreted into urine and faeces, and this proportion increases as the margin of mineral supply over mineral requirement increases. This chapter addresses the main consequences of excessive livestock mineral supplementation on environmental pollution, the toxic effects on susceptible species when exposed to mineral enriched-slurries, and the pros and cons for human health of consuming animal products with a high mineral content.

INTRODUCTION

Trace minerals are required for the normal functioning of basically all biochemical processes in the body. Providing adequate amounts of trace minerals to meet animal requirements is critical to maximizing productivity and health in animals. In the past, the primary concern was to avoid inadequate mineral supplies on production, so in the last thirty years mineral supplements have been routinely incorporated into concentrate feed at high concentrations, even when proven not strictly necessary. A significant proportion of minerals that are given to livestock are excreted into urine and faeces, and this proportion increases as the margin of mineral supply over mineral requirement increases (Suttle, 2010).

The livestock sector has a primary and growing role in agriculture economy. It is a major provider of livelihoods for the larger part of the world's poor people. Driven by growing populations and incomes, the increase in demand for animal products will be stronger than for most food items: global production of meat is projected to more than double from 229 million tonnes in 1999/2001 to 465 in 2050, and that of milk to increase from 580 to 1043 million tonnes (FAO, 2006). In this context, intensive production systems have been designed to achieve very high productivity at relatively low cost. Animals are maintained at very controlled conditions, including nutrition, to ensure minimum fatalities and increased productivity. From the point of view of mineral nutrition in order to avoid mineral deficiencies, mineral supplements are incorporated into concentrate feed in doses largely (up to 10-fold; Tables 1-3) exceeding the physiological requirements; this is possible because there are fairly wide "safety margins" to avoid mineral toxicity, even though toxic effects can appear when given at supra-optimal concentrations. However, this growth of the intensive livestock sector stresses many ecosystems and contributes to global environmental problems. Excessive mineral supplies pollute the environment and the chosen or default policy of mineral nutrition on a given farm leaves behind a "mineral footprint" (Suttle, 2010).

Table 1. Mineral requirements for growing pigs and ranges used in the industry (mg/kg)

Minerals	Requirements[a]			Industry ranges[b]	
	Swine Liveweight (kg)			Swine Liveweight (kg)	
	20-50	50-80	80-120	20-50	50-120
Cu	4	3.5	3	10-25	10-25
I	0.14	0.14	0.14	0.2-0.4	0.2-0.4
Fe	60	50	40	100-200	100-170
Mn	2	2	2	5-30	5-30
Se	0.15	0.15	0.15	0.30	0.30
Zn	60	50	50	100-200	100-200

[a] NRC, 1998. Pigs allowed feed *ad libitum* (90%) dry matter.
[b] Adapted from Zinpro Corp.

So, the future of the livestock-environment interface will be shaped by how to resolve the balance of two competing demands: for animal food products on the one hand and for the environmental services on the other (Gerber and Steinfeld, 2008).

Besides the environment, another important point is how excessive mineral supplies affect both the livestock itself and humans. In example, it is well established that bovine chronic copper toxicity is of growing concern in Europe due to the tonnes of copper that are annually dispersed around farms in policies unlikely to benefit herd health (Livesey, 2002; Miranda et al., 2010; Suttle, 2010; García-Vaquero et al., 2011a) and associated with the use of enriched-copper pig-slurry (López-Alonso et al., 2000a, 2006; Blanco-Penedo et al., 2006; Miranda et al., 2006). In the same line, excessive mineral supplementation in animal diets, are far from being able to achieve significant nutritional benefits (enriched-mineral products) for human populations which could be a risk for consumers because of the increase of toxic metal residues.

Table 2. Mineral requirements for beef cattle and ranges used in the industry (mg/kg)

Minerals	Requirement growing and finishing cattle[a]	Maximum tolerable concentration[a]	Industry ranges Feedlot growing/finishing[b]
Cr	-	-	0-1
Co	0.10	10	0.2-1.5
Cu	10	100	10-30
I	0.50	50	0.5-0.8
Fe	50	1000	0-50
Mn	20	1000	40-100
Se	0.10	2	0.1-0.3
Zn	30	500	40-120

[a] NRC, 2000.
[b] Adapted from Zinpro Corp.

Implications of Mineral Supplementation on Environmental Pollution

Ensuring environmental sustainability is one of the Millennium Development Goals to which FAO and its Members are committed. The conservation, improvement and sustainable use of natural resources, including land, water, forest, fisheries and genetic resources for food and agriculture is one of the three global goals of FAO's Strategic Framework between 2000 and 2015 (Gerber and Steinfeld, 2008).

In recent years, there has been an increase in the public concern about environmental damage instigated by intensive animal feeding operations. The European Community and other authorities like FAO have set maximum permitted levels for mineral concentrations in foodstuffs in order to protect livestock, the consumer and/or environment (EC, 2006). These follow similar precautionary principles to those adopted by producers.

For essential minerals, the authorities set maximum supplementation levels at concentrations several times greater than minimum requirements (see Tables 1-3). Manufacturers are clearly drawn by these maximum limits rather than animal need, raising mineral concentrations in feeds far above the actual needs of livestock. In other words, the main existence of these maximum permitted levels which are to protect the environment, has the opposite effect in practice, and leads to unnecessarily large "mineral footprints" (Suttle, 2010).

Excess mineral output into the environment can be reduced by feeding minerals at levels that meet the best estimates of maximum individual requirements, which means it is necessary to make changes in current commercial and legislative practice (Suttle, 2010).

Table 3. Mineral requirements for dairy cattle and ranges used in the industry

Minerals	Requirement Lactating cows [a]	Requirement Dry cows [b]	Maximum tolerable concentration [c]	Industry ranges Lactation (early) [d]	Industry ranges Dry (early) [d]
Ca (%)	0.67 (76.5 g/day)	0.44 (18.1 g/day)	1 %	>0.80	0.65-1
P (%)	0.36 (68.8 g/day)	0.22 (19.9 g/day)	1 %	>0.40	0.36-0.40
Mg (%)	0.2	0.11		>0.30	0.30
S (%)	0.2	0.2	0.4 %	0.20-0.25	0.20-0.25
Co (mg/kg)	0.11	0.11	10	0.50-1.0	0.51-1.0
Cu (mg/kg)	11	12	40	14-16	14-16
I (mg/kg)	0.44	0.4	10	0.9-1.0	0.9-1.0
Fe (mg/kg)	17	13	1000	0-30	0-30
Mn (mg/kg)	13	16	1000	50-70	50-70
Se (mg/kg)	0.3	0.3	2	0.3	0.3
Zn (mg/kg)	52	21	500	75-85	75-85

[a] Holstein cow: body weight (BW)= 680 kg, body condition score (BCS)= 3, 65 months of age, milk fat= 3.5%, milk true protein= 3.0%, milk lactose= 4.8%, days in milk= 90, milk production= 45 kg (NRC, 2001).

[b] Holstein cow: mature BW without conceptus= 680 kg, BCS= 3.3, 57 months of age, 240 days pregnant, calf weigth= 45 kg, gaining 0.67 kg/day with conceptus (NRC, 2001).

[c] NRC, 2001.

[d] Adapted from Zinpro Corp.

It has been demonstrated that the removal of all supplementary trace minerals from a fattening ration for pigs has only minor effects on performance and carcass quality, and not all are negative (Shelton et al., 2004). For this reason, there is a greater need for animal's nutritionists to develop more efficient approaches toward diet formulation that result in minimal nutrient output into the surrounding environment. Probably the most outstanding species of environmental concern are pig, poultry, and cattle and the most important minerals copper, zinc and phosphorus.

COPPER AND ZINC

In swine and poultry industry, copper and zinc are widely used at high levels as growth promoters. In pigs, dietary concentrations of 150–250 mg copper sulphate/kg and 2500–3000 mg zinc sulphate/kg (more than 25 times the minimum requirements) have been shown to stimulate growth without exposing animals to any risk of poisoning (Brumm, 1998; Poulsen, 1998). Pigs excrete approximately 80–95% of copper and zinc dietary supplements (Brumm, 1998) producing metal-enriched manures (Jondreville et al., 2003). Pig slurries typically contain high levels of copper (ca. 360 mg copper/kg dry matter (Nicholson et al., 1999), and when spread over the agricultural soils, it leads to increased soil copper concentrations (Poulsen, 1998; Bengtsson et al., 2003), toxic effects in plants and microorganisms (Coppenet et al., 1993) and other livestock species, as ruminants that are very sensitive to chronic hepatic copper accumulation (López-Alonso et al., 2000a). In a study of copper and zinc in the livers of cattle in NW Spain a positive correlation was discovered between liver copper and the density of pig-rearing units (Lopez-Alonso et al., 2000a), and cattle grazing pastures spread with pig slurry have a hepatic copper concentration above the normally accepted ''safe'' values (Blanco-Penedo et al., 2006; Lopez-Alonso et al., 2006; Miranda et al., 2006). Zinc levels are also very high in pig slurries (ca. 500 mg zinc /kg dry matter; Nicholson et al., 1999); However, this metal does not appear to pose a risk for ruminants reared on pastures amended with pig slurry due to this element being closely regulated by homeostatic mechanisms. In fact, in the above mentioned survey in Spain (Lopez-Alonso et al., 2000a) no evidence of zinc accumulation in cattle was ever found.

Great emphasis has been put on strategies for lowering the excretion of copper and zinc (Poulsen and Carlson, 2008). In Europe, the common practice of adding 250 mg/kg of copper dry matter (DM) to all pigs' rations have been changed, imposing maximum permitted levels of 170 and 25 mg/kg DM in the growing and fattening stages, respectively (EC, 2003). A similar problem arises if dietary zinc is used at high levels to reduce the incidence of diarrhea in early weaned pigs. In fact, there are new strategies to reduce this excretion, as for example to assess of the requirements of zinc and copper in weaned piglets in relation to diet composition and phytase activity (Revy et al., 2006). Another approach is to use organic copper and zinc sources instead of inorganic ones that are supposed to have a lower bioavailability, but results are not clear (Poulsen and Carlson, 2008).

A controversial point is the use of chelated minerals to reduce the outputs in the environment; that feeding hyperavailable mineral sources at reduced dietary concentrations can maintain performance in both nutritional and supranutritional contexts, while reducing the dispersal of mineral in animal wastes (Suttle, 2010). But, there is little or no evidence that this is true, in fact Veum et al. (2004) have compared fecal copper excretion by pigs given copper sulfate at growth-stimulatory levels to that of pigs given chelated copper at lower levels; the predictable result is lower fecal copper excretion with chelated copper, but the results indicates no difference in availability and that similar environmental protection could be achieved by reducing the permitted level of copper sulphate supplementation. Similarly, it has been suggested that the pharmacological benefits of adding 2–3 g/kg zinc DM as zinc oxide on the performance of early weaned, scouring piglets could be attained with less environmental impact by feeding less zinc in a chelated form (Mullan and D'Souza, 2005); however, fecal zinc excretion is determined by dietary zinc concentration rather than source.

PHOSPHORUS

Among the major minerals, the problem posed by over-feeding phosphorus has long been recognized (Joenbloed and Lenis, 1998) and is declared as one of the main pollutants of soil and waters. Dairy producers in the United States feed 0.45 to 0.50 % dietary phosphorus. This is in excess of recommendations by NRC (2001) and in excess of the needs of lactating cows (Wu et al., 2001; Valk et al., 2002), even when it is well established that the recommendations of supplementation are generally excessive (Suttle, 2010). The problem is accentuated by the over-feeding of calcium, and the fact that phosphorus is excreted in highly labile forms even on dairy farms (Toor et al., 2005; McDowell et al., 2008). Not only are dairy cattle over-fed phosphorus, the same occurring in all classes of ruminant that receive complementary feeds (Suttle, 2010) and in other species as pigs (NRC, 1998), broilers (NRC, 1994; Lukić et al., 2009) and turkeys (NRC, 1994; Godwin et al., 2005).

Phosphorus losses from livestock farms are primarily in the faeces. Fecal phosphorus increases as dietary phosphorus increases (Wu et al., 2000; Odongo et al., 2007) and can account for significant phosphorus loading to bodies of surface water, depending on the watershed. Increasing dietary phosphorus levels with phosphorus minerals not only leads to greater concentrations of total fecal phosphorus, but more importantly, increases the amount of water-soluble phosphorus, which is the most susceptible to loss into the environment (Dou et al., 2002).

From an environmental point of view, an efficient use of dietary phosphorus by dairy cows is important to minimize fecal phosphorus output and, as a result, phosphorus losses to the environment in quantities that exceed plant growth requirements (Kebreab et al., 2008).

Several strategies to reduce the impact of phosphorus on the environment as genetic manipulations of plants, microbres and animals have been developed. Biotechnology has been applied by improving the use of plant phytate phosphorus by genetic engineering (Suttle, 2010). Pigs have been genetically modified to express microbal phytase in their saliva and these breeds are able to utilize plant phytate phosphorus better than conventional pigs (Golovan et al., 2001).

More recently, poultry has been modified to overexpress a recombinant avian phosphatase with high activity towards plant phytate phosphorus (Cho et al., 2006); However, results are not clear and extensive commercial application remains a distant prospect (Suttle, 2010).

Easier and cheaper is the dietary manipulation. In pigs and poultry for example, replacement of corn with a low phosphorus; feeding grain by-products of the distillery which are of much higher phosphorus than the grains themselves; reducing calcium supplements to the minimum; bleeding corn and soybean meal with cereals and by-products rich in phytase activity; adding water prior to feeding to stimulate endogenous phytase activity; substituting animal protein for soybean protein, since its phosphorus is three times more available, etc. (Suttle, 2010).

In dairy cattle, it has been suggested that dairy farmers in Ontario (Canada) could save themselves CAN $20 per cow and spare the environment 1.3 kt phosphorus per year by reducing the phosphorus concentration in feed from 0.41 to 0.35% DM without impairing cow health of productivity (Kebreab et al., 2008).

In fact, Odongo et al. (2007) suggested that the current forages grown in Ontario contain adequate amounts of phosphorus to sustain milk production in mature dairy cows without the need for mineral phosphorus supplementation or without affecting animal productivity.

EFFECTS OF MINERAL SUPPLEMENTATION ON ANIMAL HEALTH

As previously indicated, mineral supplementation in livestock reared under intensive production is considered relatively safe because the high degree of dietary standardization and the relatively high security margins for most minerals and in most domestic species. For this reason episodes of intoxication associated to mineral supplementation are relatively rare, at least in causing a change in the mineral bioavailability or an error in the mineral supplement formulation occurs.

In spite of this, especial consideration should be given to hepatic chronic accumulation in ruminants, this is because the high difficulty to establish copper requirements in these species depending in other dietary components and the low capacity of these animal species for billiary excretion. In addition, the high heavy metal content of mineral supplements compared to other feedstuffs and their interactions with some essential minerals (especially for cadmium) can lead to toxic effects on animals (López-Alonso et al., 2002a, 2004; Blanco-Penedo et al., 2006).

CHRONIC COPPER TOXICITY IN RUMINANTS

It is well known that copper is essential for life processes, as a cofactor for many vital cuproenzymes, but is extremely toxic in excess (Mercer, 2001); because of its dual role, all living organisms have developed highly specialized homeostatic mechanisms to recruit, deliver and eliminate copper, and to neutralize its toxic effects.

Within the domestic animals, there are marked variations in their tolerance to increased levels of dietary copper (Howell and Gooneratne, 1987). Whereas pigs are highly tolerant to dietary copper and it can be supplemented at concentrations as high as 250 mg/kg DM as growth promoters (well above the physiological needs: 3-4 mg/kg DM; NRC; 1998), sheep are the most susceptible species to chronic copper toxicity and supplementation is restricted to 15 mg/kg DM (EC, 2003). It has been demonstrated that sheep, as opposed to pigs, have a limited capacity to accumulate copper bound to metallothionein in their livers (Howell and Gooneratne, 1987) and a very limited capacity to increase biliary copper excretion in response to increased copper intake (Bremner, 1998). Hepatic subcellular distribution studies have demonstrated that the lysosomes copper storage capacity saturates at lower total hepatic copper loads (Gooneratne et al., 1979), leading to chronic copper concentrations in the nucleus and cytosol at lower exposures that are toxic for other animal species. It is also well known that there are strong genetic differences in susceptibility to copper accumulation and certain breeds of sheep have been classified as tolerant or resistant to copper toxicity (Suttle, 2010) and as a result it has been possible to improve the resistance of sheep to copper deficiency and excess by appreciating cross breeding and selection programmes (Suttle 1986; Suttle et al., 2002) depending on copper availability on the diet.

Within the other ruminant species, cattle were traditionally thought to be relatively tolerant to copper accumulation and reports of copper poisoning were, until recently, somewhat rare. In fact, copper deficiency in cattle is a rather common disorder worldwide and cattle diets are regularly supplemented with high copper concentrations (up to 35 mg/kg dry matter (DM), the maximum level of copper supplementation for cattle established by the European Union; EC, 2003), is well above general physiological requirements (10 mg/kg DM; NRC, 2000). The relatively wide range for copper supplementation is because in cattle, and in ruminants in general, copper nutritional requirements do not depend exclusively on dietary copper concentrations, but are highly dependent on the copper availability (copper from cereals in concentrate feeds is more available than from forage, NRC 2001) and the presence of other dietary elements that can influence copper absorption and metabolism, mainly molybdenum, sulphur, iron and zinc (Kendall et al., 2001; NRC, 2001; Suttle, 2010). In fact, elevated copper supplementation has, in some cases, been justified in view of the interference of copper with its antagonists (Kendall et al., 2001). Even more so, some particular conditions should be considered when establishing copper dietary requirements, for example in calves the very high copper absorption efficiency form milk or in weaned heifers the administration of coccidiostat with copper supplementation, ignoring the fact that the medication would defaunate the rumen and greatly increase copper absorbability (Suttle, 2010).

It is well documented that chronic copper poisoning in ruminants is a two-stage process. During the first stage (prehaemolytic phase) copper accumulates in the liver without clinical signs of disease, and only when the liver storage capacity is overloaded (usually following a stressful event of some sort), is when an important hepatic damage occurs, leading to the release of high amounts of copper into the blood stream causing a rupture of erythrocytes (haemolytic phase) (Howell and Gooneratne, 1987). The clinically silent phase of the disease makes diagnosis difficult until the animal suffers the haemolytic crisis.

In fact a recent study noted that herds with episodes of clinical copper toxicity, the clinical cases represent only a small proportion of the actual cases of copper poisoning (Laven et al., 2004; López-Alonso et al., 2006). Identification of animals in this silent prehaemolitic stage of copper accumulation can be considered very important to avoid not only economic losses due to subsequent severe disease or death, but also to avoid subclinical disease.

In recent years an increase in the number of episodes of copper toxicity has been reported in cattle reared under intensive systems (Bidewell et al., 2000; VLA, 2001; Laven et al., 2004), even at liver concentrations well below those regarded as toxic in the bibliography (Perrin et al., 1990; Gummow, 1996).

In most cases cattle toxicity is associated with an excessive copper intake in the ration, as well as with changes in the type and bioavailability of dietary copper supplements (Galey et al., 1991; Stefen et al., 1997; Laven et al., 2004), which has highlighted the need to redefine copper needs of cattle reared on intensive conditions (Garcia Vaquero et al., 2011a).

It has also been reported that dietary supplements leading to copper accumulation in the liver at concentrations only slightly above normal (of around 125 mg/kg wet weight) showed negative effects on animal performance in terms of reduced feed intake and average daily gain (Engle and Spears, 2000).

Very recently, results found by our research group have demonstrated that under the conditions of cattle raised in intensive systems in NW Spain, which are similar to those in many European countries, copper supplementation (15 mg/kg DM) routinely leads to a hepatic copper accumulation in the liver exceeding the adequate range in most animals (90%), half of them showing hepatic copper concentrations associated with toxicity (García-Vaquero et al., 2011a); copper supplementation did not suppose significant change in the copper storage capacity and the antioxidant defensive system evaluated by MT and SOD expression, but with a significant and important increase of oxidative damage measured by iNOS and NITT (García-Vaquero et al., 2012).

Liver copper concentrations that seemingly could be associated with subclinical chronic copper toxicity, as described in the above mentioned experiments in cattle, have been described in many countries where copper supplements are given well above requirements (Hadrich, 1996; Jilg et al., 1997) and could have a real impact on animal performance and consequently the animal production.

Toxic Metal Toxicity in Farm Animals Associated to Mineral Supplementation

Mineral supplements generally contain trace residues of toxic metals, especially of cadmium and lead (EFSA 2004a, b). In fact, mineral supplements are considered one of the feedstuffs with the highest toxic metal concentrations, and the maximum admissible concentrations established by the European Union (EC, 2002) are higher than for most other feedstuffs.

However, its contribution to the total dietary intake is difficult to evaluate since mineral supplements and premixes may be added at different rates according to manufacturer's instruction.

EFSA has evaluated the risk of the main toxic metals present in the animal feed on the animal health and it was considered negligible for mineral supplements, if their content is within the maximum admissible levels established by the European Union.

The only exception seems to be cadmium, not only because its bioacumulative properties can accumulate at very high concentrations in the liver, but especially in the kidney of adult animals causing adverse effects including renal function impairment, hypertension, disturbance of trace mineral metabolism and acute degenerative damage in the intestinal villi. Minimum toxic levels or maximum safe dietary concentrations cannot be estimated with any precision, since cadmium deposition is significantly influenced by dietary interactions with copper, zinc, iron and calcium; thus, in some cases, concentrations of cadmium as low as 1 mg/kg in the diet did in fact induce adverse effect in animals (EFSA, 2004a).

In addition, a significant increase of cadmium accumulation in tissues can be observed when high dietary supplements of copper and zinc are used in the ration (EFSA, cadmium). This is due to these metals having similar chemical properties, being able to induce and compete for the binding sites of metallothioneins. This is especially relevant for pigs, where copper and zinc are used at very high concentrations as growth promoters and have a very high capacity to induce metallothioneins (Henry et al., 1994); for example: cadmium residues in the liver and kidney were double in animals receiving 200 mg copper/kg DM for three months compared to the controls (Rothe et al., 1992).

EFFECTS ON HUMAN NUTRITION—*PROS AND CONS*

The life-sustaining properties of micronutrients in foods consumed by humans and the associated possibilities for enhancing both functional and marked value by changing methods of food production (Welch and Graham, 2004; Blanco Penedo et al., 2010) have opened a new area of research on animal nutrition. Lucrative markets for trace element supplements for humans raise concerns that modern diets are inadequate and some claimed benefits of these supplements supposedly improve fertility, vitality and resistance to infections (Suttle, 2010).

Animal products (including meat, offal, as well as milk and eggs) are considered important sources of minerals for humans because of their high content of minerals compared to other foods (López-Alonso et al., 2000b, 2002b, 2007). From an animal nutrition point of view, there has been the temptation to over-provide minerals in animal diets to get mineral-enriched animal products. However, it is not always possible to enhance the levels of oligoelements; this is because there are important physiological homeostatic mechanisms that control the absorption-tissue accumulation and excretion of minerals by the organism. The reason for this regulation is the fact that essential trace elements are also toxic to cellular metabolism when present at uncontrolled quantities. The homeostatic control of essential trace element flux through the body is mediated by down-regulation of absorption and/or stimulation of excretion (urine, faecal excretion from endogenous sources) (Kirchgessner, 1993) and it becomes active once the dietary supply of essential trace elements turns from deficient to sufficient amounts, hence it has been largely used to determine the requirement of essential trace minerals. Consequently, determination of mineral concentrations in sensitive tissues (e.g. zinc blood plasma or bone concentrations), activities of metalloenzymes (e.g. alkaline phospatase or superoxide dismutase) as well as contents in animal products (e.g. zinc concentrations in eggs or milk) reflect an absence or presence of homeostatic regulation and thus indicate reliably the point of dietary supplies matching the metabolic requirement.

Homeostatic control at the site of intestine absorption is known for the most important trace elements as zinc, copper and iron (Kirchgessner, 1993) even though important interactions and competitions are known when one of these elements is over-supplied in relation to the others. On the contrary, for other essential trace elements, like selenium, iodine and cobalt, their homeostatic control is mediated mainly by urinary excretion. These elements are absorbed from the intestinal tract at high rates without homeostatic control over a wide range of supply from deficient to considerable excess. As homeostatic elimination from the body becomes active after absorption the respective elements intrude into the organism in a manner proportional to intake, which explains why bodily fluids and tissue contents of these elements usually continue to raise at dietary supplies exceeding requirements, being those elements with a higher potential to achieve enriched-mineral animal products.

Particular interest surrounds the ability of selenium at supranutritional levels to reduce susceptibility to carcinogens (Rayman, 2005; Irons et al., 2006).The supplementation of livestock rations with selenium raises milk selenium in dairy cows (Juniper et al., 2006; Phipps et al., 2008), egg selenium in hens (Utterback et al., 2005) and muscle selenium in lambs (Juniper *et al.*, 2008) selenomethionine, such as selenium yeast, being more effective than inorganic selenium to increase animal products selenium concentrations. Most of the additional selenium in milk (Phipps *et al.*, 2008) and muscle (Juniper *et al.*, 2008) is present as selenomethionine and the partition of selenium in the egg shifts from the yolk to the albumen when selenium yeast is used instead of selenite (Leeson *et al.*, 2008). The success of selenomethionine is attributable to its intact absorption and resistance to degradation but availability for glutathione peroxidase synthesis in the supplemented animal is relatively poor, and turnover rate in muscle is slow (Juniper *et al.*, 2008). Less interest has been put on iodine supplementation to achieve iodine-enriched animal products; however iodine content in the milk can be improved feeding supranutritional levels of iodine to the dairy cow, because there is a linear relationship between dietary and milk iodine (Suttle, 2010).

Because of its great importance to human nutrition, mineral content in meat has been largely evaluated (López Alonso et al., 2000b, 2002b, 2007). It is important to consider that in situations of an adequate mineral status, whereas mineral concentration in edible viscera (liver and kidney) can be variable (even 2-fold) depending on the degree of mineral storage capacity, essential trace element concentrations in muscle are irrespective of the mineral status of the animal and could possibly be related to their own particular muscular metabolism (García-Vaquero et al., 2011b). The most active and less fat containing muscles showed in general the highest essential and the lowest non-essential trace element accumulation.

The production system largely conditions the mineral content of animal products. In a study carried out in NW Spain comparing trace element concentrations in animal tissues in beef cattle from different production systems, it was observed that animals that received high levels of mineral supplementation (typically from intensive feedlot systems) showed generally higher levels of trace elements in the liver and kidney than those from organic systems (where mineral supplementation is not-allowed or highly restricted). This was especially true for copper in the liver where animals from intensive systems accumulate hepatic concentrations 2-3 times the physiological levels. On the contrary, the muscle essential trace element concentrations in these animals were within adequate physiological ranges (in spite of animals from organic systems showing copper, cobalt and selenium deficiencies considering their hepatic concentrations). There were no significant correlations in elements concentrations between muscle and liver and kidney (organ concentrations that better reflect exposure), except for cobalt where there was a positive association (Blanco Penedo et al., 2009).

During the interval between the slaughter of an animal and the cooking, autolytic changes take place that affect the taste, smell and visual appearance of the product. The process is essentially one of oxidative deterioration and consequences such as change of colour (e.g. accumulation of metmyoglobin) (Ryu et al., 2005), exudation ('watery pork') and taint (Bou et al., 2005) can have a negative effect on retailer and consumer valuation of meat products. Manufacturers of mineral supplements with potential antioxidant properties, including zinc chelates and selenium-Met, have been quick to search for evidence that feeding them to livestock enhances the sensory value of meat, measuring everything from pH and shear strength to 'tastiness' in blindfolded consumers in the hope of positive outcomes similar to those found with vitamin E supplements. However, these trials have been conducted with selenium given in combination with vitamin E (Ryu et al., 2005). Improvements in storage characteristics have generally been small (Bou et al., 2005). For example, a supranutritional inorganic selenium supplement produces only minor improvements in the oxidative stability of chicken meat (Ryu et al., 2005). An organic selenium supplement that raised the selenium level in a pig ration from 0.18 (high) to 0.48 mg kg−1 DM (very high) reduced 'drip loss' significantly and more effectively than inorganic selenium (Mateo et al., 2007), but whether the reduction of 0.8% in drip loss, only one-third of the total loss, justifies 'medication' of the pork is questionable.

Finally, a relevant point related to the high or even excessive use of mineral supplementation in farm animals is the high toxic metal concentration in mineral supplements. As indicated, it is well assumed that in relatively unpolluted rural areas, the feed mineral supplements are the main source of toxic metals for animals. However, because of its low rate of inclusion in the concentrate feed, their contribution to the total dietary metal exposure is low and toxic metal residues in meat and meat products (mainly offal) are generally below the maximum admissible levels established by the EU (for review see López-Alonso et al., 2000b, 2007) even though they can pose a real risk for accumulative elements like cadmium, where a low concentration in the diet can lead to high levels of residues (above the maximum admissible levels) in offal of adult animals. The risk of high cadmium residues is also high in offal of young animals receiving very high copper and zinc dietary supplementation, especially in pigs where they are used as growth promoter. To avoid animal products with a high toxic metal content reaching the food chain, monitoring programs are regularly conducted in most developed countries, and depending on the toxic metal residues offal from adult animals, they are generally not recommended for human consumption.

Conclusion

In intensive production systems mineral supplements are routinely incorporated into concentrate feed, even when not necessary, at concentrations largely exceeding the physiological requirements, thus being responsible of unnecessarily large "mineral footprints". New research is needed to better adjust mineral supplementation to physiological needs. Special attention should be given to copper because of its widespread use on animal nutrition (especially in pigs as a growth promoter) and the great susceptibility of ruminants to chronic copper toxicity.

References

Bengtsson, H; Öborn, I; Jonsson, S; Nilsson, I; Andersson, A. Field balances of some mineral nutrients and trace elements in organic and conventional dairy farming—a case study at Öjebyn, Sweden. *European Journal of Agronomy*, 2003, 20, 101–116.

Bidewell, CA; David, GP; Livesey, CT. Copper toxicity in cattle. *The Veterinary Record*, 2000, 147, 399–400.

Blanco-Penedo, I; Cruz, JM; López-Alonso, M; Miranda, M; Castillo, C; Hernández, J; Benedito, JL. Influence of copper status on the accumulation of toxic and essential metals in cattle. *Environment International*, 2006, 32, 901-906.

Blanco-Penedo, I; Shore, RF; Miranda, M; Benedito, JL; López-Alonso, M. Factors affecting trace element status in calves in NW Spain. *Livestock Science*, 2009, 123(2/3), 198–208.

Blanco-Penedo, I; López-Alonso, M; Miranda, M; Hernández, J; Prieto, F; Shore, RF. Non-essential and essential trace element concentrations in meat from cattle reared under organic, intensive or conventional production systems. *Food Additives and Contaminants,* 2010, 27 (1), 36–42.

Bou, R; Guardiola, P; Barroeta, AC; Codony, R. Effect of dietary fat sources and zinc and selenium supplements on the consumption and consumer acceptability of chicken meat. *Poultry Science*, 2005, 84, 1129–1140.

Bremner, I. Manifestations of copper excess. *American Journal of Clinical Nutrition,* 1998, 67, 1069S–1073S.

Brumm, MC. Sources of manure: Swine. In: Hatfield JL, Stewart BA, editors. *Animal waste utilization: Effective use of manure as a soil resource*. Michigan: Ann Arbor Press; 1998; 49–64.

Cho, J; Choi, K; Darden, T; Reynolds, PR; Petite, JN; Shears, SB. Avian multiple inositol polyphosphate phosphatase is an active phytase that can be engineered to help ameliorate the planet's 'phosphate crisis'. *Journal of Biotechnology*, 2006, 1, 248–259.

Coppenet, M; Golven, J; Simon, J.C; Le Corre, L; Le Roy, M. Évolution chimique des sols en explotations d'élevage intensif: Exemple du Finistère. *Agronomie*, 1993, 13, 77–83.

Dou, Z; Knowlton, KF; Kohn, RA; Wu, Z; Satter, LD; Zhang, G; Toth, JD; Ferguson JD. Phosphorus characteristics of dairy feces affected by diets. *Journal of Environmental Quality*, 2002, 31, 2058–2065.

EC. Directive 2002/32/EC of the European parliament and of the Council of 7 May 2002 on undesirable substances in animal feed. *Journal of the European Commission* OJ, 2002, L140/10, 30.05.2002, p.12.

EC. Commission Regulation No. 1334/2003 of the European Parliament and the Council. *Official Journal of the European Commission* L187/11–L187/15, Annex, 2003.

EC. Commission Regulation No 1881/2006 of 19 December 2006 setting maximum levels for certain contaminants in foodstuffs. *Journal of the European Commission* OJ, 2006, L364/5, 20.12.2006, p.20.

EFSA (European Food Safety Authority). Opinion on the Scientific Panel on Contaminants in the Food Chain on a request from the Commission related to cadmium as undesirable substance in animal feed. Adopted on 2 June 2004. *The EFSA Journal*, 2004a, 72, 1-24.

EFSA (European Food Safety Authority). Opinion on the Scientific Panel on Contaminants in the Food Chain on a request from the Commission related to lead as undesirable substance in animal feed. Adopted on 2 June 2004. *The EFSA Journal*, 2004b, 71, 1-20.

Engle, TE; Spears, JW. Effects of dietary copper concentration and source on performance and copper status of growing and finishing steers. *Journal of Animal Science*, 2000, 78, 2446–2451.

FAO. World Agriculture: towards 2030/2050, Interim Report. Rome, Italy, 2006.

Galey, FD; Maas, J; Tronstad, RJ; Woods, LW; Jonson, BJ; Litlefield, ES; Wallstrum, R; Dorius, LC. Copper toxicosis in two herds of beef calves following injection with copper disodium edetate. *Journal of Veterinary Diagnostic and Investigation*, 1991, 3, 260–263.

García-Vaquero, M; Miranda, M; López-Alonso, M; Castillo, C; Benedito, J.L. Evaluation of the need of copper supplementation in intensively reared beef cattle. *Livestock Science*, 2011a, 137, 273–277.

García-Vaquero, M; Miranda, M; Benedito, JL; Blanco-Penedo, I; López-Alonso, M. Effect of type of muscle and Cu supplementation on trace element concentrations in cattle meat. *Food and Chemical Toxicology,* 2011b, 49, 1443-1449.

García-Vaquero, M; López-Alonso, M; Benedito, JL; Miranda, M. Histochemistry evaluation of the oxidative stress and the antioxidant status in Cu supplemented cattle. *Animal,* 2012, *In press*.

Gerber, PJ; Steinfeld, H. Worldwide growth of animal production and environmental consequences. In: Schlegel P, Durosoy S, Jongbloed A, editors. *Trace elements in animal production systems*. The Netherlands: Wageningen Academic Publishers; 2008; 21-32.

Godwin, JL; Grimes, JL; Christensen, VL; Wineland, MJ. Effect of dietary phosphorus and phytase levels on the reproductive performance of large white turkey breeder hens. *Poultry Science*, 2005, 84, 485–493.

Golovan, SP; Meidinger, RG; Ajakaiye, A; Cottrill, M; Wiederkehr, MZ; Barney, DJ; Plante, C; Pollard, JW;Fan, MZ. Pigs expressing salivary phytase produce low-phosphorus manure. *Nature Biotechnology*, 2001, 19, 741–745.

Gooneratne, SR; Howell, J.McC; Gawthorne, J. Intracellular distribution of copper in the liver of normal and copper loaded sheep. *Research in Veterinary Science*, 1979, 27, 30–37.

Gummow, B. Experimentally induced chronic copper toxicity in cattle. *Onderstepoort Journal of Veterinary Research*, 1996, 63, 277–288.

Hadrich, J. High amounts of copper in calf's livers. Recent data and estimation of potential health hazards. *Deutsche Lebensmittel-Rundschau*, 1996, 92, 103–113.

Henry, RB; Liu, J; Choudhuri, S; Klaassen CD. Species variation in hepatic metallothionein. *Toxicology Letters*, 1994, 74, 23-33.

Howell, J.McC; Gooneratne, RS. The pathology of copper toxicity in animals. In: Howell J.McC, Gawthorne JM, editors. Copper in Animals and Man, vol. II. Florida: CRC Press; 1987; 53–78.

Irons, R; Carlson, BA; Hatfield, DL; Davies, CD. Both selenoproteins and low molecular weight selenocompounds reduce colon cancer risk in mice with genetically impaired selenoprotein expression. *Journal of Nutrition*, 2006, 136, 1311–1317.

Jilg, T; Unglaub, W; Eckstein, B. Influence of copper supplementation in milk replacers on the copper concentration of calf livers. *Fleischwirtschaft*, 1997, 77, 559–562.

Joengbloed, AW; Lenis, NP. Environmental concerns about animal manure. *Journal of Animal Science,* 1998, 76, 2641–2648.

Jondreville, C; Revy, PS; Dourmad, JY. Dietary means to better control the environmental impact of copper and zinc by pigs weaning to slaughter. *Livestock Production Science*, 2003, 84, 147-156.

Juniper, DT; Phipps, RH; Jones, AK; Bertin, G. Selenium supplementation of lactating dairy cows: effect on selenium concentration in blood, milk, urine and feces. *Journal of Dairy Science*, 2006, 89, 3544–3551.

Juniper, DT; Phipps, RH; Ramos-Morales, E; Bertin, G. Selenium persistency and speciation in the tissues of lambs following the withdrawal of a high-dose selenium-enriched yeast. *Animal*, 2008, 2, 375–380.

Kebreab, E; Odongo, NE; McBride, BW; Hanigan, MD; France, J. Phosphorus utilization and environmental pollution and economic implications of reducing phosphorus pollution from Ontario dairy farms. *Journal of Dairy Science*, 2008, 91, 241–246.

Kendall, NR; Illingworth, DV; Telfer, SB. Copper responsive infertility in British cattle: the use of a blood caeruloplasmin to copper ratio in determining a requirement for copper supplementation. In: Diskin MG, editor. *Fertility in the High-Producing Dairy Cow.* Edinburgh: Occasional Publication 26 (2), British Society of Animal Science; 2001; 429-432.

Kirchgessner, M. Homeostasis and homeorhesis in trace element metabolism. In: Anke M, Meissner D, Mills CF, editors. Trace Element in Man and Animal, TEMA 8. Gerstorf Germany: Verlag Media Touristik; 1993; 4-21.

Laven, RA; Livesey, CT; Offer, NW; Fountain, D. Apparent subclinical hepatopathy due to excess copper intake in lactating Holstein cattle. *The Veterinary Record*, 2004, 155, 120–121.

Leeson, S; Namkung, H; Caston, L; Durosoy, S; Schlegel, P. Comparison of selenium levels and sources and dietary fat quality in diets of broiler breeders and laying hens. *Poultry Science*, 2008, 87, 2605–2612.

Livesey, CT; Bidewell, CA; Crawshaw, TR; David, GP. Investigation of copper poisoning in adult cows by the Veterinary Laboratories Agency. *Cattle Practice*, 2002, 10, 289–294.

López-Alonso, M; Benedito, JL; Miranda, M; Castillo, C; Hernández, J; Shore, RF. The effect of pig farming on copper and zinc accumulation in cattle in Galicia (North-Western Spain). *The Veterinary Journal*, 2000a, 160, 259-266.

López-Alonso, M; Benedito, JL; Miranda, M; Castillo, C; Hernández, J; Shore, RF. Toxic and trace elements in liver, kidney and meat from cattle slaughtered in Galicia (NW Spain). *Food Additives and Contaminants,* 2000b, 17, 447–457.

López-Alonso, M; Benedito, JL; Miranda, M; Castillo, C; Hernández, J; Shore, RF. Interactions between toxic and essential trace metals in cattle from a region with low levels of pollution. *Archives of Environmental Contamination and Toxicology*, 2002a, 42, 165-172.

López-Alonso, M; Benedito, JL; Miranda, M; Castillo, C; Hernández, J; Shore, RF. Contribution of cattle products to dietary intake of trace and toxic elements in Galicia, Spain. *Food Additives and Contaminants*, 2002b, 19, 533–541.

López-Alonso, M; Montaña, FP; Miranda, M; Castillo, C; Hernández, J; Benedito, JL. Interactions between toxic (As, Cd, Hg and Pb) and nutritional essential (Ca, Co, Cr, Fe, Mn, Mo, Ni, Se, Zn) elements in the tissues of cattle from NW Spain. *Biometals,* 2004, *17, 389-397.*

López-Alonso, M; Crespo, A; Miranda, M; Castillo, C; Hernández, J; Benedito, JL. Assessment of some blood parameters as potential markers of hepatic copper accumulation in cattle. *Journal Veterinary Diagnostic Investigation*, 2006, 18, 71–75.

Lopez-Alonso, M; Miranda, M; Castillo, C; Hernandez, J; García-Vaquero, M; Benedito, JL. Toxic and essential metals in liver, kidney and muscle of pigs at slaughter in Galicia, north-west Spain. *Food Additives and Contaminants*, 2007, 24, 943-954.

Lukić, M; Pavlovski, Z; Škrbić, Z. Mineral nutrition of modern poultry genotypes. *Biotechnology in Animal Husbandry*, 2009, 25 (5-6), 399-409.

Mateo, RD; Spalholz, JE; Elder, R; Yoon, I; Kim, SW. Efficacy of dietary selenium sources on growth and carcass characteristics of growing-finishing pigs fed diets containing high endogenous selenium. *Journal of Animal Science*, 2007, 85, 1177–1183.

McDowell, RW; Dou, Z; Toth, JD; Cade-Menun, BJ; Kleinman, PJA; Soder, K; Saporito, L. A comparison of the phosphorus speciation and potential bioavailability in feed and feces of different dairy herds using 31P nuclear magnetic resonance spectroscopy. *Journal of Environmental Quality*, 2008, 37, 741–752.

Mercer, JFB. The molecular basis of copper-transport diseases. *Trends in Molecular Medicine*, 2001, 7, 64–69.

Miranda, M; Cruz, JM; López-Alonso, M; Benedito, JL. Variations in liver and blood copper concentrations in young beef cattle raised in north-west Spain: Associations with breed, sex, age and season. *Animal Science*, 2006, 82, 253–258.

Miranda, M; Gutiérrez, B; Benedito, J.L; Blanco-Penedo, I; García-Vaquero, M; López-Alonso, M. Influence of breed on blood and tissue copper status in growing and finishing steers fed diets supplemented with copper. *Archives of Animal Nutrition*, 2010, 64, 98–110.

Mullan, B; D'Souza, D. The role of organic minerals in modern pig production. In: Taylor-Pickard JA, Tucker LA, editors. *Redefining Mineral Nutrition*. Nottingham, UK: Nottingham University Press; 2005; 89–117.

National Research Council (NRC). Nutrient Requirements of Beef Cattle. 7th. Revised Edition. Washington: National Academy of Sciences, National Academic Press; 2000.

National Research Council (NRC). Nutrient Requirements of Dairy Cattle. 7th. Revised Edition. Washington: National Academy of Sciences, National Academic Press; 2001.

National Research Council (NRC). Nutrient Requirements of Poultry. 9th. Revised Edition. Washington: National Academy of Sciences, National Academic Press; 1994.

National Research Council (NRC). Nutrient Requirements of Swine. 10th. Revised Edition. Washington: National Academy of Sciences, National Academic Press; 1998.

Nicholson, FA; Chambers, BJ; Williams, JR; Unwin, RJ. Heavy metal content of livestock feeds and animal manures in England and Wales. *Bioresource Technology*, 1999, 70, 23–31.

Odongo, NE; McKnight, D; Koekkoek, A; Fisher, JW; Sharpe, P; Kebreab, E; France, J; McBride, BW. Long-term effects of feeding diets without mineral phosphorus supplementation on the performance and phosphorus excretion in high-yielding dairy cows. *Canadian Journal of Animal Science*, 2007, 87(4), 639-646.

Perrin, DJ; Schiefer, B; Blakley, BR. Chronic copper toxicity in a dairy herd. *Canadian Veterinary Journal*, 1990, 31, 629–632.

Phipps, R.H; Grandison, AS; Jones, AK; Juniper, DT; Ramos-Morales, E; Bertin, G. Selenium supplementation of lactating dairy cows: effects on milk production and total selenium content and speciation in blood, milk and cheese. *Animal*, 2008, 2, 1610–1618.

Poulsen, HD Zinc and copper as feed additives, growth factors or unwanted environmental factors. *Journal of Animal and Feed Sciences*, 1998, 7, 135–42.

Poulsen, HD; Carlson, D. Zinc and copper for piglets - how do high dietary levels of these minerals function?. In: Schlegel P, Durosoy S, Jongbloed A, editors. *Trace elements in animal production systems*. The Netherlands: Wageningen Academic Publishers; 2008; 151-160.

Rayman, MP. Selenium in cancer prevention: a review of the evidence. *Proceedings of the Nutrition Society*, 2005, 64, 527–542.

Revy, PS; Jondreville, C; Dourmad, JY; Nys, Y. Assessment of dietary zinc requirement of weaned piglets fed diets with or without microbial phytase. *Journal Animal Physiology an Animal Nutrition*, 2006, 90, 50-59.

Rothe, S; Gropp, J; Weiser, H; Rambeck, WA. Influence of vitamin-c and zinc on copper-induced increased cadmium retention in pig. *Zeitschrift Fur Ernahrungswissenschaft*, 1994, 33, 61-67.

Ryu, YC; Rhee, MS; Lee, KM; Kim, BC. Effects of different levels of dietary supplemental selenium on performance, lipid oxidation and color stability of broiler chicks. *Poultry Science*, 2005, 84, 809–815.

Shelton, JL; Southern, LL; LeMieux, FM; Bidner, TD; Page, TG. Effects of microbial phytase, low calcium and phosphorus, and removing the dietary trace mineral premix on carcass traits, pork quality, plasma metabolites, and tissue mineral content in growing-finishing pigs. *Journal of Animal Science*, 2004, 82, 2630–2639.

Steffen, DJ; Carlson, MP; Casper, HH. Copper toxicosis in suckling beef calves associated with improper administration of copper oxide boluses. *Journal of Veterinary Diagnostic and Investigation*, 1997, 9, 443–446.

Suttle, NF. Mineral Nutrition of Livestock. 4th ed. UK: CAB International; 2010.

Suttle, NF. Copper deficiency in ruminants: recent developments. *The Veterinary Record*, 1986, 119, 519-522.

Suttle, NF; Lewis, RM; Small, JNW. Effects of breed and family on rate of copper accretion in the liver of purebred Charollais, Suffolk and Texel lambs. *Animal Science*, 2002, 75, 295-302.

Toor, GS; Cade-Munn, BJ; Sims, JT. Establishing a linkage between phosphorus forms in dairy diets, faeces and manures. *Journal of Environmental Quality*, 2005, 34, 1380–1391.

Utterback, PL; Parsons, CM; Yoon, I; Butler, J. Effect of supplementing selenium yeast in diets of laying hens on egg selenium content. *Poultry Science*, 2005, 84, 1900–1901.

Valk, H; Sebek, LBJ; Beyen, AC. Influence of phosphorus intake on excretion and blood plasma and saliva concentrations of phosphorus in dairy cows *Journal of Dairy Science*, 2002, 85, 2642–2649.

Veum, TL; Carlson, MS; Bollinger, DW; Ellersieck, MR. Copper proteinate in weanling pig diets for enhancing growth performance and reducing faecal copper excretion compared with copper sulfate. *Journal of Animal Science*, 2004, 82, 1062–1070.

VLA Surveillance Report. July sees an increased incidence of copper poisoning in cattle. *The Veterinary Record*, 2001, 149, 257–260.

Welch, RM; Graham, RD. Breeding for micronutrients in staple food crops. *Journal of Experimental Botany*, 2004, 55, 353–364.

Wu, Z; Satter, LD; Blohowiak, AJ; Stauffacher, RH; Wilson, JH. Milk production, estimated phosphorus excretion, and bone characteristics of dairy cows fed different amounts of phosphorus for two or three years. *Journal of Dairy Science*, 2001, 84, 1738–1748.

Wu, Z; Satter, LD; Sojo, R. Milk production, reproductive performance, and fecal excretion of phosphorus by dairy cows fed three amounts of phosphorus. *Journal of Dairy Science*, 2000, 83, 1028–1041.

In: Trace Elements ISBN: 978-1-62081-376-8
Editors: Diego A. De Leon and Paloma Raquel Aragon © 2012 Nova Science Publishers, Inc.

Chapter 5

Distribution and Possible Sources of Trace Elements in Estuarine Sediments: Comparative Study among Estuaries of the Basque Country and other Parts of the World

Ainara Gredilla, Silvia Fdez-Ortiz de Vallejuelo, Azibar Rodriguez, Julen Bustamante, Gorka Arana, Alberto de Diego* and Juan Manuel Madariaga

Department of Analytical Chemistry, University of the Basque Country (UPV/EHU), Bilbao, Basque Country, Spain

Abstract

Although trace elements are natural constituents of the environment, several of them present high toxicity, long persistency and quick accumulation in living organisms. For these reasons identification of trace element contamination sources and evaluation of their environmental impact are essential. Trace element pollution in estuaries has been considered a wide world problem for a long time. The principal pollution sources are atmospheric deposition (volcanic activity, combustion of fossil fuels, etc), land run-off (agricultural, mines, etc), direct industrial and urban inputs, fisheries and recreational uses. Once in water, trace elements are mainly adsorbed on suspended particle matter or colloids and deposited by flocculation in sediments, where they present different chemical forms. Thus sediments are important deposits of trace elements and hence an important tool to assess metal pollution in estuaries. Bibliographic data on the concentration of trace elements (As, Cd, Cr, Cu, Fe, Mn, Ni, Pb and Zn) measured in sediments from different estuaries of the world has been collected, summarized and compared with those experimentally obtained in two of the most important estuaries in the Basque Country (Urdaibai and Nerbioi-Ibaizabal). In general, it can be said that the concentration of trace element in sediments from the Nerbioi-Ibaizabal and Urdaibai

* Tel.:+34 94 601 5550, E-mail address: alberto.dediego@ehu.es.

estuaries are not alarming in comparison with other estuaries. Different tools have been used to identify diffuse and point sources in both estuaries.

Keywords: Trace elements, pollution monitoring, estuaries, sediments

1. Distribution and Possible Sources of Trace Elements

Pollution by trace elements has been considered a world level problem since the early seventies.

It is well known that some trace elements present harmful consequences on the environment and on human health, due principally to their high toxicity, long persistency and quick accumulation in organisms (El-Moselhy and Gabal, 2004). The identification of contamination sources and the evaluation of their environmental impact in aquatic environments are, therefore, necessary.

In pristine aquatic environments, the concentration of trace metals in sediment and water is directly related to the composition of the basin. Unfortunately, this is rarely true, mainly due to the influence of human activity. At the moment, trace metals are one of the most important pollutants that affect transitional waters. They enter the estuaries as a result of natural processes or processes derived from human activity. The geochemistry of the basin, together with wind-blown dusts, highly influences the first ones. The processes related to human activity result from external sources as direct atmospheric deposition, rivers or direct discharges.

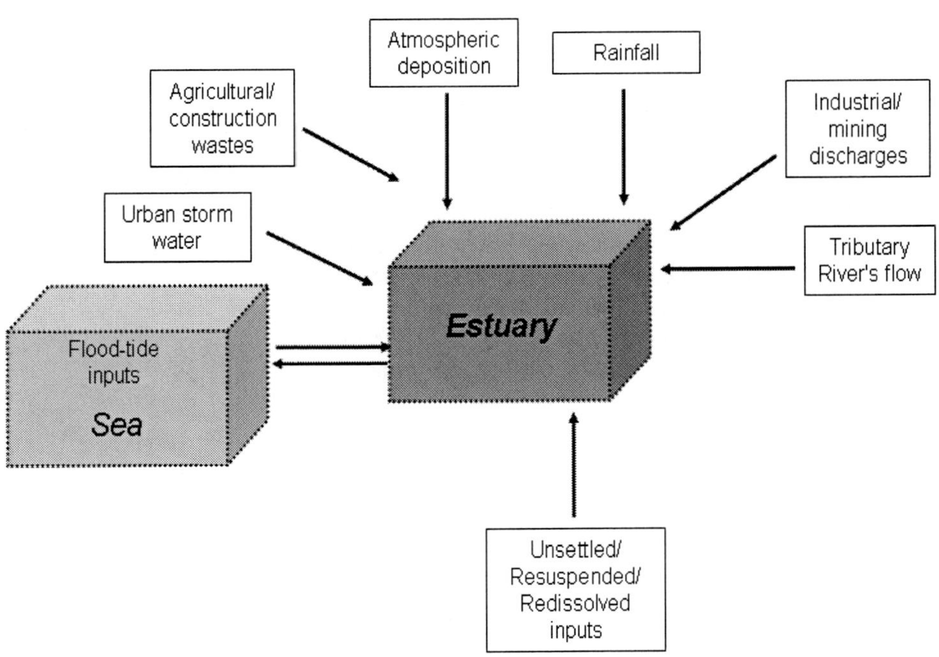

Figure 1. Overall scheme of pollution sources that commonly affect an estuary.

The atmospheric deposition is mainly due to volatile compounds as Cd, Zn and Pb, which could be transported as gas or ashes hanged in the air and finally deposited in the water or in the sediment (Demirak et al., 2006). The direct inputs are usually the consequence of agricultural (Pd and Cd), urban (Al, Cu, Fe, Pb, Ni and Zn), mining (As, Cu, Cd, Pb and Mn) and industrial (Cd, Cu, Fe, Pb, Mn, Ni and Zn) waste products (Scancar et al., 2000). A scheme of the principal pollution sources of trace elements in estuaries is shown in Figure 1.

Once in estuary, the future of trace elements is clearly conditioned by competitive processes between water and sediment, such as particle/solute interactions, flocculation and coagulation, sedimentation, resuspension from the sediment bed, and/or diffusion of metals from the interstitial water to the water column.

Depending on the process, trace elements present a conservative or a non-conservative behaviour, that is why transitional areas are commonly known as "filters" for both suspended particulate matter and trace elements. Trace elements entering the estuaries do not remain much time as water-soluble forms. About the 90% of trace elements that goes into the aquatic system is transferred to the sediment by adsorption on the surface of suspended particulate matter (< 0.45 μm) (Dai et al., 1995) or colloids (< 0.2 μm) (Censi et al., 2006).

Throughout the process an important part of water is trapped into the sediment, forming the interstitial water. Trace elements tend to equilibrium between the sediment and interstitial water, which will be certainly affected by the physico-chemical properties of the environment. For this reason the metal content measured in superficial sediments is commonly considered a good indicator of the water current quality (Spencer et al., 2003).

In unpolluted sediments trace elements are mainly limited to silicate forms and primary minerals with limited mobility. In polluted areas, however, metals tend to associate with organic substances (humic and fulvic acids) or aluminosilicates. They can also be adsorbed into iron and manganese oxide and/or hydroxides and precipitate as carbonates, sulphates or hydroxides. Moreover, as they easily join the fine grain of the sediments, the 1-3 cm depth surface layer is the most representative one regarding metal pollution (Szava-Kovats, 2008). Although in polluted areas metals are usually associated to these phases with high mobility, they are not usually available to organisms. However punctual or seasonal changes in the physico-chemical properties of the water column (pH, salinity, redox potential or dissolved oxygen) may reverse this situation and become sediments a potential risky pollution source.

2. Trace Elements in Estuarine Sediments

As estuaries connect land, rivers and coastal waters they have been always considered highly productive water bodies and, consequently, they have been intensely exploited. Besides their ecological importance, they have also been notable economic resources, as centres of industrial and urban activities, among other uses. It has to be also taken into account that estuaries provide convenient routes for transportation and conduits for disposal of wastewaters. For all of these reasons they have been considered the most productive coastal environments.

The geochemistry of estuaries is quiet complex. They present changeable hydrodynamics with strong spatial and temporal variations. These variations affect the physico-chemical properties of estuaries, producing vertical and lateral gradients in salinity, pH, suspended particulate matter, dissolved oxygen and both organic and inorganic nutrients.

Salinity is an important environmental factor that plays a fundamental role in estuaries, determining the reactions that take place on them. The geochemical reactivity of trace metals, for example, is commonly attributed to changes in metal adsorption-desorption equilibrium along the salinity gradient of estuaries.

Basing on the salinity gradient, the behaviour of trace elements in these estuaries can be conservative (if the trace element concentration varies linearly with salinity) or non-conservative (if the contrary is observed) (Audry et al., 2007). Nitrogen (as nitrates), phosphorous (as phosphates) and silicon associated to trace metals formed the most important non-conservative elements.

A non-conservative behaviour could be the result to different aspects as the following: i) the variability of composition and fluxes of freshwater over time, ii) the variability observed in water residence time in the estuary and/or iii) the different chemical composition of the flow which can change depending on the season. The change in ion ratio is directly related to all these aspects and determines the chemical processes around the salinity gradient. It has been demonstrated that in sea water diluted by fresh water the principal modification in the behaviour of the chemical components takes place at the critical salinity of 5-8 PSU (practical salinity units) (Telesh and Khlebovich, 2010).

When trace elements enter the estuary they take part in different processes. Flocculation is one that highly depends on the salinity gradient. This mechanism is common in areas of low salinity and takes place when the organic colloids from the river water meet principal cations coming from the sea. This joint is maximum at a salinity of 20 PSU. According to the affinity degree between trace elements and marine anions, and between the first ones and colloidal forms of the humic compounds, the flocculation rate is of the following order: Fe (95%) > Cu, Ni (40%) > Mn (25-45%) > Al (20%) > Co (10%) > Cd (5%) (Mendiguchía Martínez, 2005).

Sorption is another important mechanism of trace elements that changes along the salinity gradient in estuaries. Some studies have concluded that there is a decrease in the affinity between the trace elements and the solid phase when the salinity increases. This behaviour could be ascribed to the increase in competitiveness between the polyvalent cations of the sea water and the surface of the particles. It has been proved that as salinity increases the sorption rates decrease as follows: Cd > Ni > Cu, Pb > Zn (Mendiguchía Martínez, 2005).

In any case, approaches based exclusively on the study of biogeochemical processes of trace elements and on their concentrations do not take into account the specific geochemical characteristics of each estuary. The use of a variety of numerical indexes can help in this sense. The enrichment factor, EF (Muñoz Barbosa et al., 2010), the geoaccumulation index, I_{geo}, (Müller, 1979) and the contamination factor, CF, (Hakanson, 1980) have been commonly used with this purpose. They give us the possibility to differentiate between the natural and anthropogenic fraction of the trace element contained in the sediment, using the background value defined for the area as a reference.

3. TRACE ELEMENTS IN DIFFERENT ESTUARIES OF THE BASQUE COUNTRY

3.a. The Estuary of Urdaibai

Urdaibai is a meso-macrotidal estuary (13 Km long and 500 m wide) located in the Bay of Biscay (Basque Country) (43° 22′N, 2° 40′W). The Urdaibai estuary (also known as Mundaka or Gernika estuary) is within a natural region that was declared a Biosphere's Reserve by UNESCO in 1984 and it is considered a relatively unpolluted area (Puy-Azurmendi et al., 2006). However, it is located relatively close to several industries including metallurgic, shipyards, dyes, cutlery and manufacture (Bartolome et al., 2006). The Urdaibai region covers an area of 220 km^2 with around 45.000 habitants, most of them concentrated in Bermeo and Gernika towns (population ~ 20.000). This territory is characterized by a hydrographic basin of small streams (Oka River, and other minor effluents like Golako, Mape, Artike and Laga) that merge in a great salt marsh surrounded by high sheer cliffs (Bartolome et al., 2006).

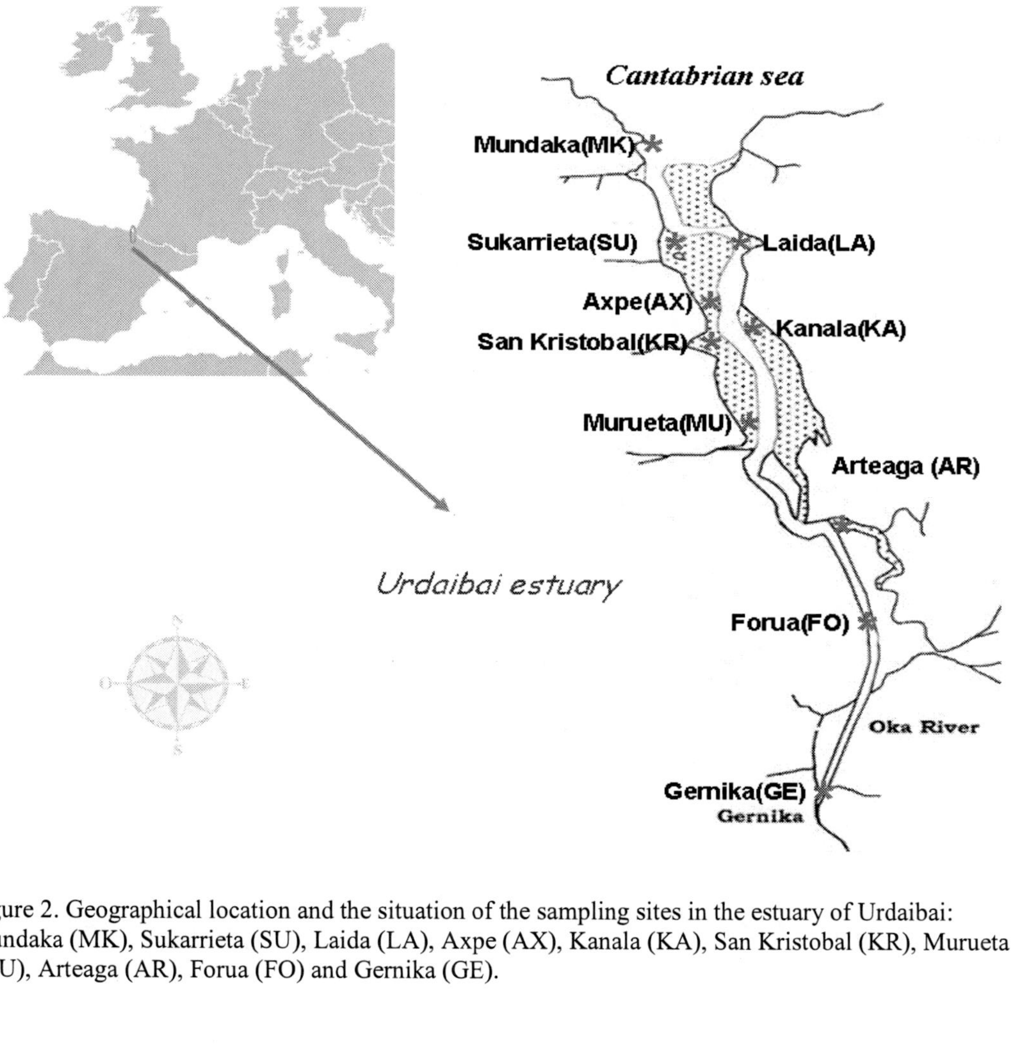

Figure 2. Geographical location and the situation of the sampling sites in the estuary of Urdaibai: Mundaka (MK), Sukarrieta (SU), Laida (LA), Axpe (AX), Kanala (KA), San Kristobal (KR), Murueta (MU), Arteaga (AR), Forua (FO) and Gernika (GE).

The outer estuary is characterized by salinity values around or higher than 31 PSU and a well mixed water column. In the intermediate section salinity values between 25 PSU and 31 PSU are common. The inner section is a narrow artificial channel (about 15 m wide) which shows a marked riverine influence and its salinity is a function of river flow and tidal condition (Orive et al., 1995). The freshwater residence time in the estuary varies between 1day and 60 days during prolonged dry periods (Orive et al., 1998).

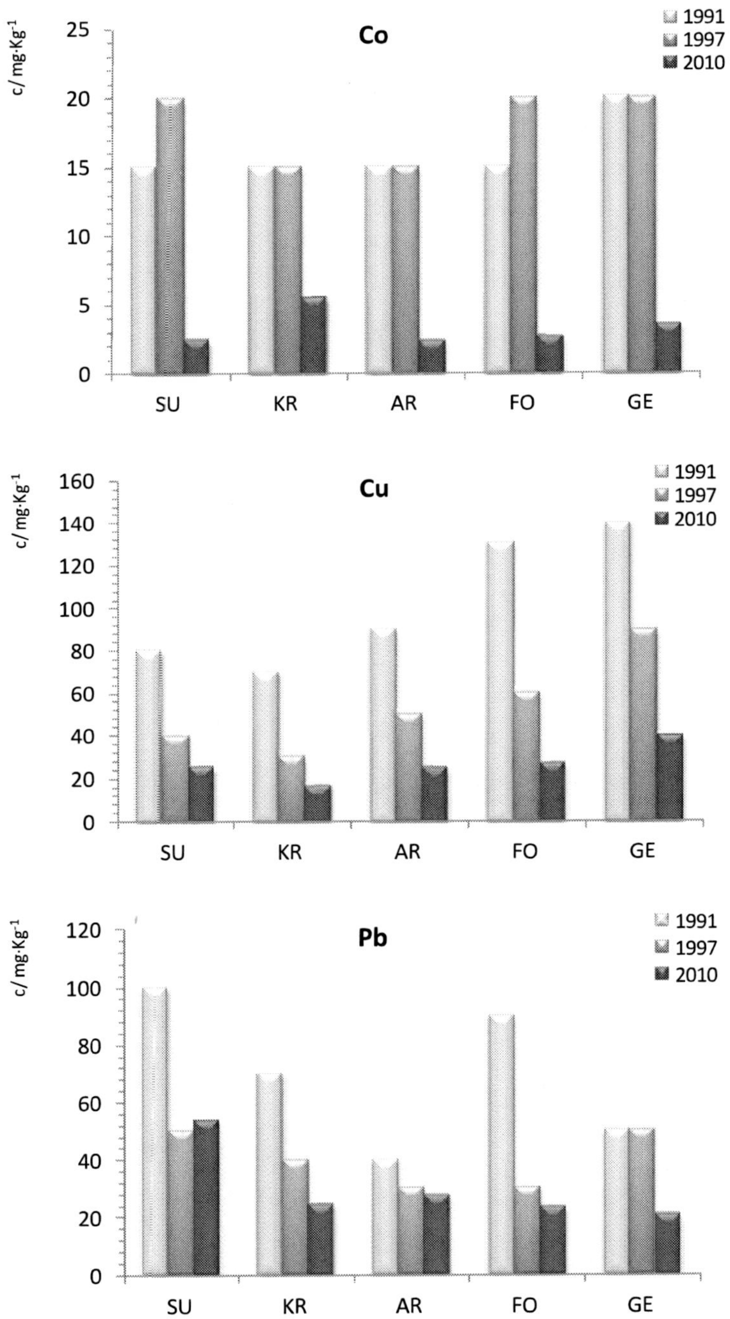

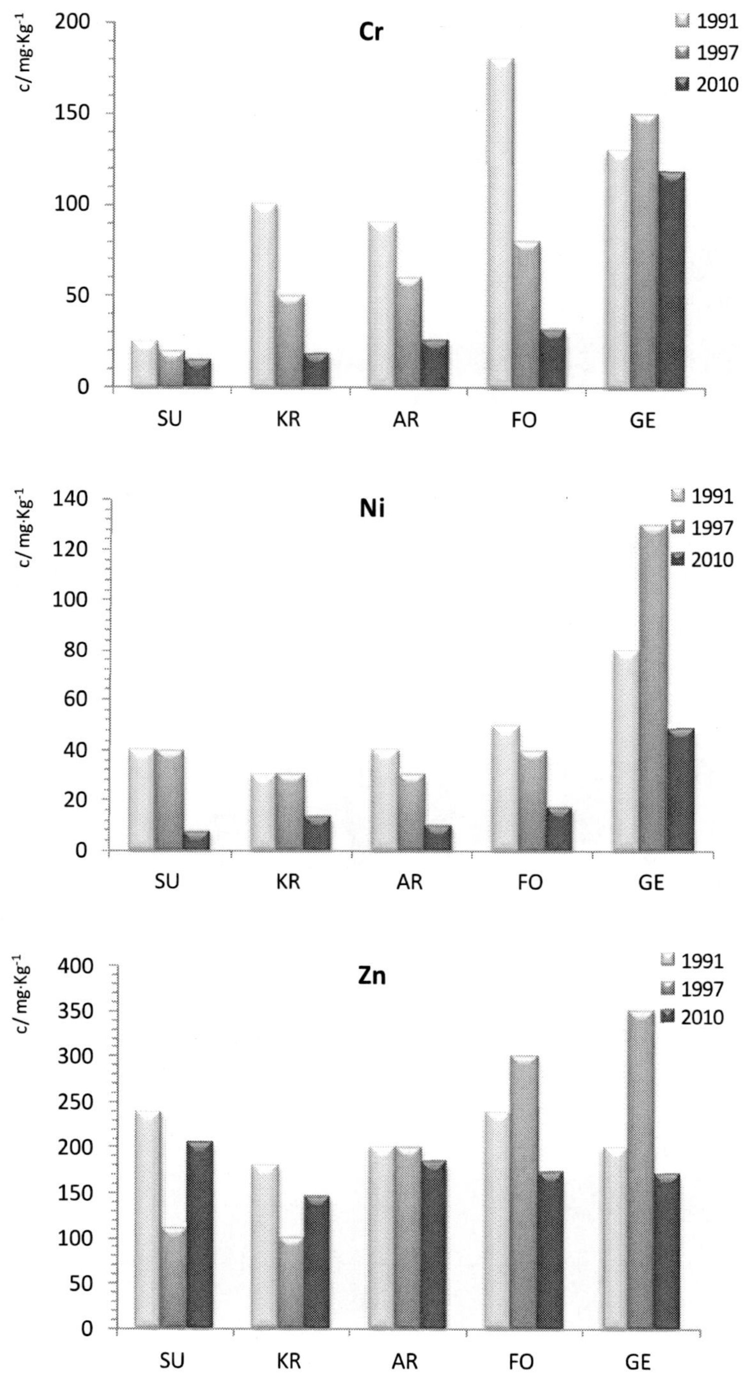

Figure 3. Concentration of trace elements ($mg{\cdot}kg^{-1}$) found in sediments collected at different points of the estuary of Urdaibai in October 1991, October 1997 (Irabien and Velasco, 1999) and March 2010.

In a preliminary study Irabien and Velasco (Irabien and Velasco, 1999), measured the concentration of Fe, Mn, Zn, Pb, Cu, Cr, Ni and Co in superficial sediments (fraction < 63 μm) collected at five sites of the estuary of the Oka River in October 1991 and October 1997. The samples were digested and measured by atomic absorption spectrometry (AAS).

In the first sampling, they found relatively high concentrations of metals in a sampling site near the town of Gernika, which the authors explained like urban and industrial waste discharged directly to the estuary. In the second sampling (1997), they found a significant decrease in some heavy metals level, especially in the case of Pb and Cu, except in the area of Gernika city for Zn, Cr and Ni.

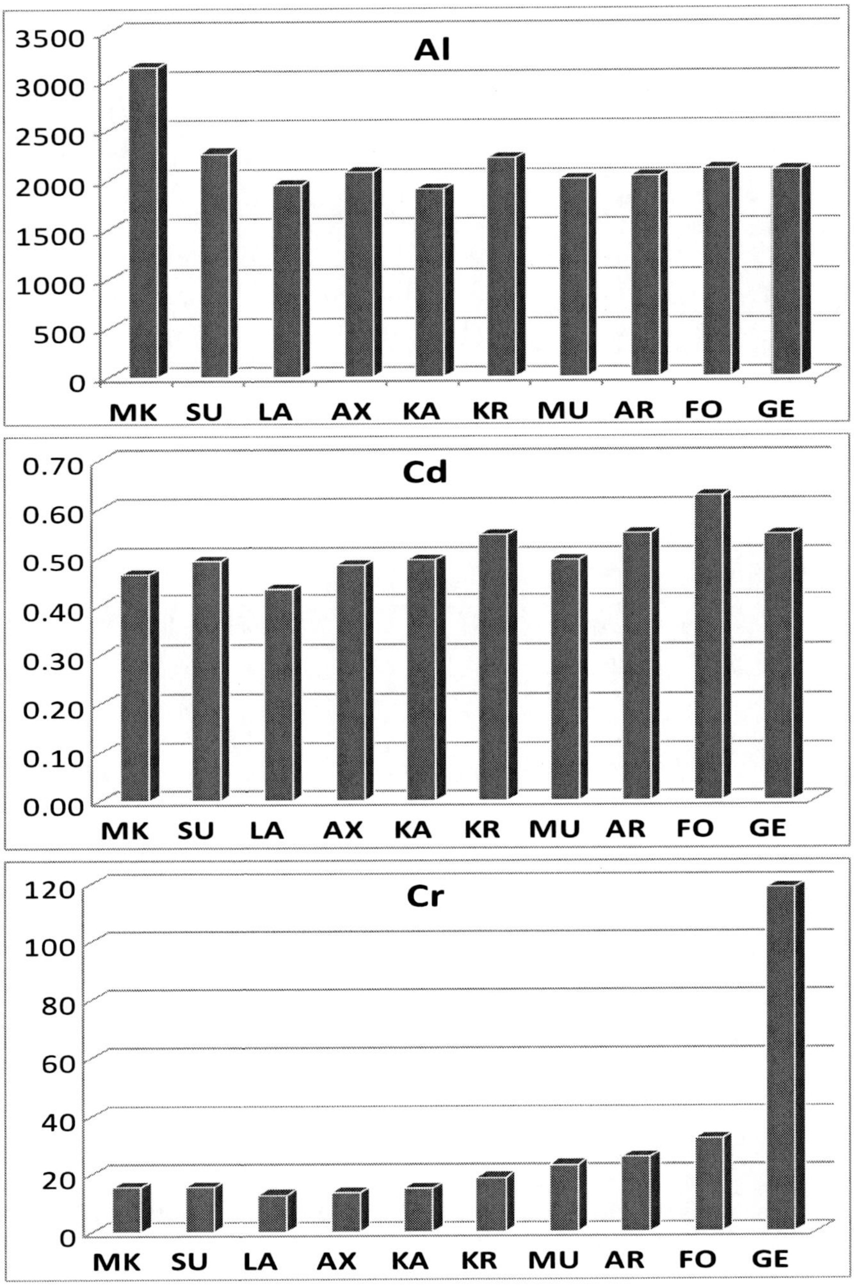

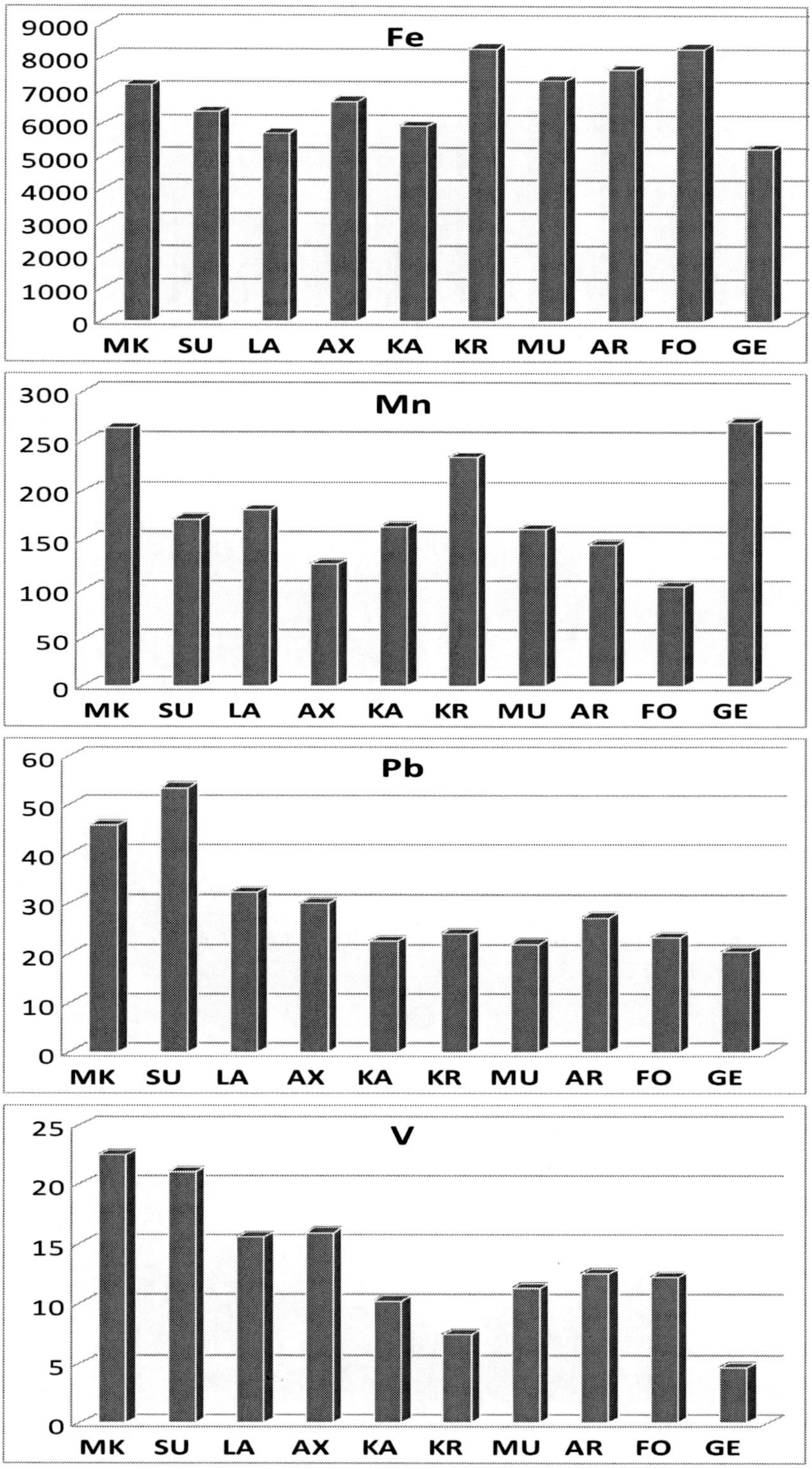

Figure 4. (Continued).

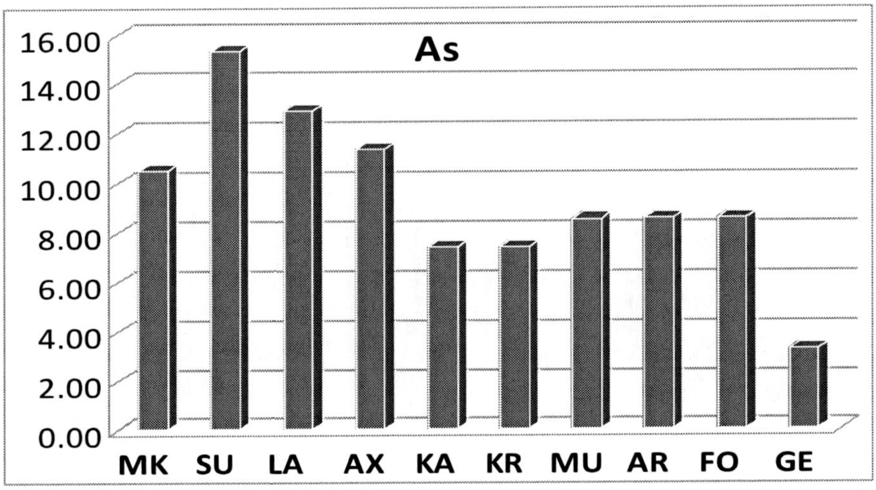
As
16.00
14.00
12.00
10.00
8.00
6.00
4.00
2.00
0.00
MK SU LA AX KA KR MU AR FO GE

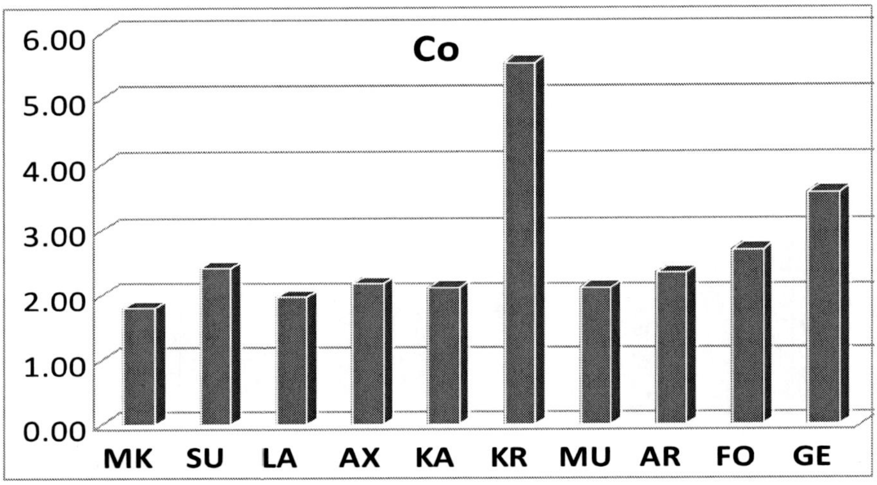
Co
6.00
5.00
4.00
3.00
2.00
1.00
0.00
MK SU LA AX KA KR MU AR FO GE

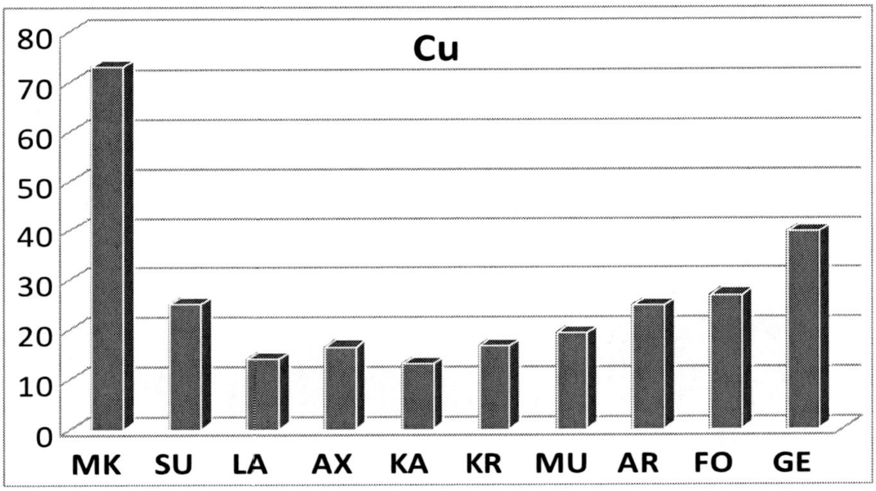
Cu
80
70
60
50
40
30
20
10
0
MK SU LA AX KA KR MU AR FO GE

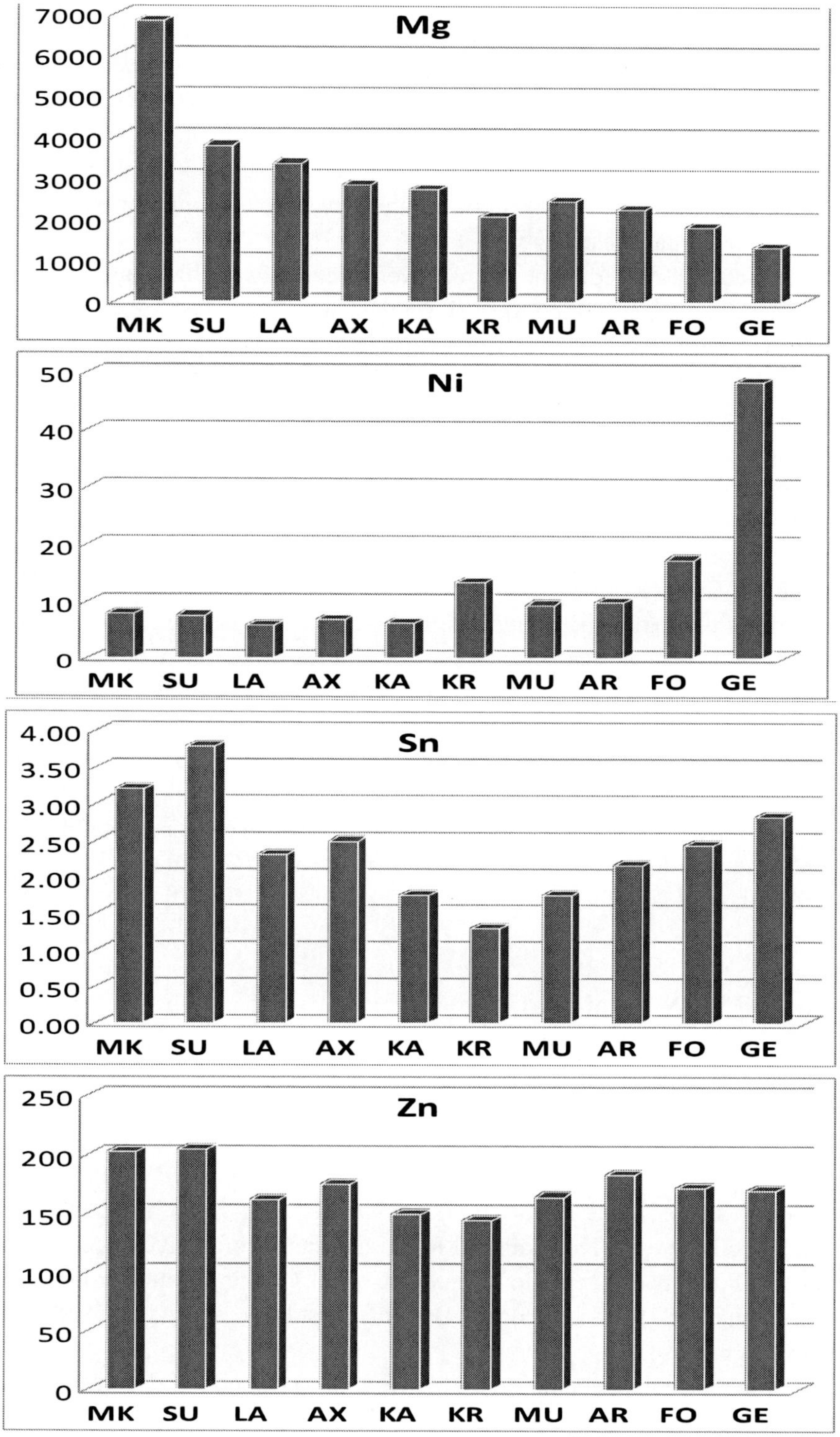

Figure 4. Concentrations of Al, As, Cd, Co, Cr, Cu, Fe, Mg, Mn, Ni, Pb, Sn, V and Zn ($mg \cdot Kg^{-1}$) measured in each sampling point of the estuary of Urdaibai in Mach 2010.

In an attempt to verify the trends observed by Irabien and Velasco, and to identify the possible sources of heavy metals into estuary, we collected surface sediments (~ 2 cm depth) at 10 sampling points of the estuary in March 2010. The samples were collected at low tide with a plastic spade and using latex gloves to avoid sample contamination. Then they were put into clean plastic containers and were finally taken to the laboratory in portable cooler boxes at 4ºC. Figure 2 shows the position of the ten selected sampling sites strategically distributed along the Urdabai estuary.

Once in the laboratory, sediment samples were frozen, lyophilized and finally sieved to assure a maximum particle size of 65 μm. All samples were kept at 4ºC until analysis.

The concentrations of Al, As, Cd, Co, Cr, Cu, Fe, Mg, Mn, Ni, Pb, Sn, V and Zn were measured by inductively coupled plasma mass spectrometry (ICP/MS) after focused assisted acid extraction of the samples (Fdez-Ortiz et al., 2009).

Five of the sampling sites selected by Irabien and Velasco are near the localizations selected in this study, e.g. Sukarrieta (SU), San Kristobal (KR), Arteaga (AR), Forua (FO) and Gernika (GE). The results obtained in those points can be observed in Figure 3. A general decreasing trend along over time was observed in the studied areas, particularly in the case of Co, Cu and Ni. For Cr in Gernika the decreasing pattern was not so clear.

Figure 4 shows the spatial distribution of the 14 elements measured in March 2010.

As it can be appreciated, the distribution of Fe and Zn is homogeneous along the estuary, but in the case of Al, As, Mg, Mn (except in Gernika), Pb and V decreasing concentration profiles can be distinguished as the distance from the mouth increases. Cd, Co, Cr, Cu (except Mundaka) and Ni show similar trends, with higher concentration in Gernika (the upper part of the estuary) than near the sea.

It would be interesting to check if the variations in time (every three months) observed in trace element concentration in sediments from other estuaries (Gredilla et al., 2012b) are also present in the estuary of Urdaibai.

3.b. Estuary of the Nerbioi-Ibaizabal River

The estuary of the Nerbioi-Ibaizabal River may be considered as the paradigm of the estuaries in the Basque Country. For many years it has been the most important industrial and commercial area in the Basque Country (mostly due to iron exploitation in local mines at the end of the 19th century).

The Nerbioi-Ibaizabal River estuary is situated in the south-east of the Bay of Biscay (43º 20′N, 3º 1′W) (see Figure 5). The Ibaizabal River and the Nerbioi River meet in Basauri and the confluence of the two rivers flows into an estuary taking water from another four tributaries: Kadagua (%27) and Galindo (%4) on the west bank, and Asua (%0.7) and Gobela (%0.3) on the east bank.

In terms of annual relative salinity, and according to the classification of the European Union (WFD, 2000), the estuary is oligohaline (between ‰0.5 and ‰5) at low tide and euhaline (between ‰30 and ‰40) at high tide. Regarding the average tidal range, it is mesotidal (between 2 m and 4 m).

The estuary valley has a surface area of 1700 km^2, its average flow rate is 30 $m^3 \cdot s^{-1}$ (Belzunce et al., 2001) and the quantity of sediment that gathers on the estuary bed is between 14.000 and 1.000.000 tonnes per year (Arias et al., 2008).

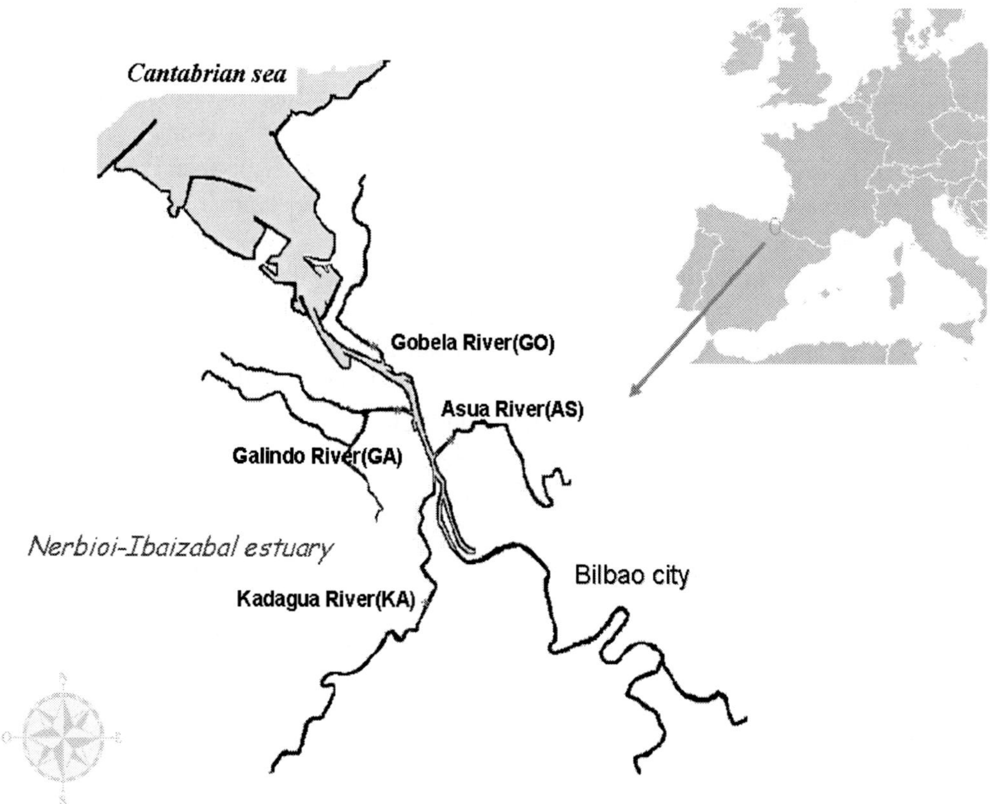

Figure 5. Geographical location of the estuary of the Nerbioi-Ibaizabal River.

This transitional aquatic area receives impacts from a population of about one million inhabitants, that is 87% of the population of Biscay and 29% of the population of the Basque Country. Additionally, constructions that affect river flow (docks, anchorages, etc.) and ports have significantly altered the river mouth. There is also a large waste water treatment plant on the north bank of the Galindo River, barely 500 m from the river mouth.

Much work has been carried out regarding the assessment of the ecological status of the estuary of the Nerbioi-Ibazabal River. A pioneering work was that published by Seebold, Labarta and Amigó in 1982 (Seebold et al., 1982), which studies the distribution of titanium and other metals in sediments of the estuary.

Long-term monitoring of sediments has allow to go into spatial and temporal variation in metal concentration in the estuary (Cearreta et al., 2002; Tueros et al., 2009; Fdez-Ortiz de Vallejuelo et al., 2010; Gredilla et al., 2012a). The biological impact of metal elements accumulated in sediment has also been studied (Leorri et al., 2008; Fdez-Ortiz de Vallejuelo et al., 2010; Fdez-Ortiz de Vallejuelo et al., 2011; Gredilla et al., 2012b). Detailed analyses of the abundance of mercury and its species (methylmercury, in particular) have also been carried out (Bartolome et al., 2006). The historical evolution of pollution in the estuary has also been followed by sediment cores that have been dated and analysed to determine the presence of benthic foraminifera and the concentration of metals (Cearreta et al., 2000; Cearreta et al., 2002; Cundy et al., 2003). Metal background values for the area have also been estimated (Rodríguez et al., 2006). Studies that aim to determine possible sources of metal in the estuary have also been carried out (Belzunce et al., 2001).

Long-term monitoring carried out by our research group (Fdez-Ortiz de Vallejuelo et al., 2010; Gredilla et al., 2012a) has allowed us to calculate some statistics on the trace element concentrations measured in sediments collected from January 2005 to October 2010. Samples were collected every three months at eight different points of the estuary. The analytical procedure followed to collect and analyse the samples was the same than that applied in the case of the sediments of the estuary of Urdaibai. The results are summarised in Figure 6 in the form of Box-Whisker plots.

The sediments with the highest content of toxic elements like As, Cd, Cr, Cu, Ni, Pb and Zn are concentrated in the surroundings of the Gobela and Galindo Rivers (the last is probably affected by the sewage treatment plant located in its banks). Sediments from the upper part of the estuary are less affected by metal pollution and the sediments of the outer part of the estuary are especially rich in Al, Mg and Mn.

In adition, possible pollution sources have also been studied. Moreover, a depth study of the geographical distribution showed by each metal along the estuary and point and diffuse pollution sources may be distinguished. The presence of Al and Mg in sediments, for example, is similar all over the estuary with a decreasing concentration profile as the distance from the mouth increases.

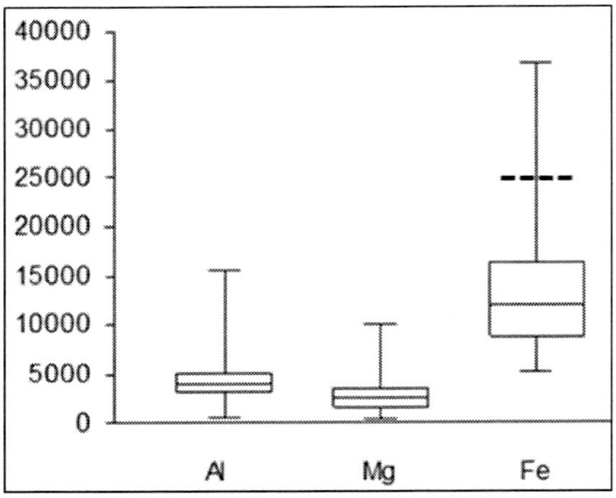

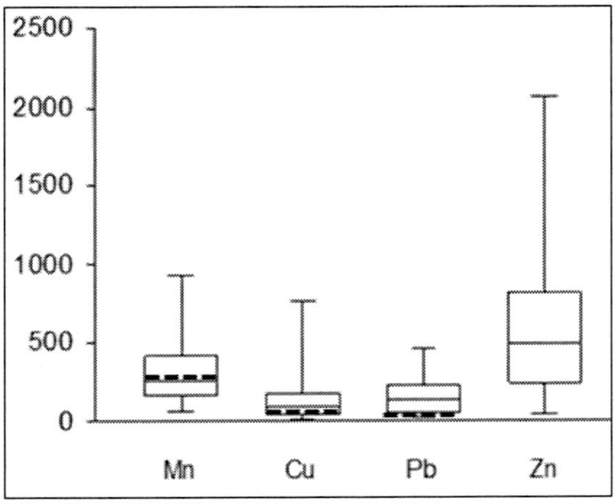

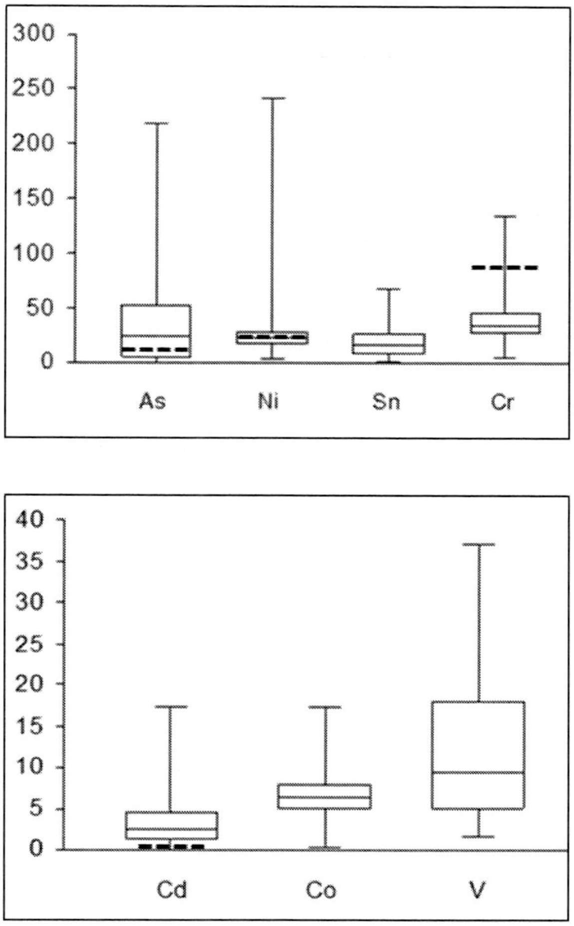

Figure 6. Box-Whisker plots of the trace element concentrations (mg·kg^{-1}) measured in sediments collected between January 2005 and October 2010. The box shows the 25th percentile and the 75th percentile, and the whiskers represent the lowest and the highest concentrations, while the line inside the box expresses the average value. The discontinuous line indicates the background value defined for each metal (Cearreta et al., 2000; Rodríguez et al., 2006).

This fact may indicate a common pollution source located in the surroundings of the Abra Bay (where the estuary flows into the ocean) or, simply a direct dependence of their concentration on the salinity of water.

Fe, Sn, Zn and Pb also present similar distribution patterns along the estuary, with high concentrations in the surroundings of Gobela and Galindo tributaries. These rivers could be defined as point pollution sources of these metals.

The area close to the Gobela River also presents high concentrations of As, Cd, Cu and V. The Co and Ni concentrations measured in the sediments do not differ significantly from site to site, except close to the Galindo River, which may be acting again as a possible pollution source for these two metals.

The case of Cr is special. Its distribution is quite uniform in the whole estuary (with the exception of the area close to the Kadagua River, where low Cr content was found) and it is the unique trace element which presets high concentrations in the sediments of the upper part of the estuary. The input of Cr to the estuary is probably of diffuse origin and may present different sources as dry or wet atmospheric deposition or urban runoff.

It should be taken into account that the studies based on the simple metal content are only valid to investigate the spatial distribution of pollution in a selected environment at a given moment. They are site- and time-specific. In order to improve this and with the objective to obtain a real idea of the pollutant fraction accumulated in sediments, the geoccumulation indexes were calculated. These coefficients give the possibility to differentiate between the natural and anthropogenic fraction of the trace element contained in the sediment, using the background value defined for the area (in Figure 6 reference values defined for the sediments of the Nerbioi-Ibaizabal estuary are included).

The geoaccumulation indexes calculated for the Nerbioi-Ibaizabal estuary allowed as to draw some conclusions on the origin of the elements considered. Although some punctual samples present high Fe, Ni and Cr content, the sediments of the Nerbioi-Ibaizabal estuary are not significantly polluted with these metals.

As, Pb, Zn and Cd on the contrary, and Ni, Cu and Zn in a lower extent, present high concentrations with probably anthropogenic origin (see background values in Figure 6).

4. Trace Elements in Sediments of Different Estuaries of the World

Detailed assessments of the distribution, storage and potential environmental impact of contaminants in sediments have been carried out in different estuaries and coastal areas of the world. Regarding our study the concentration of heavy metal in sediments of the estuary of Urdaibai is much lower than that detected in the estuary of the Nerbioi-Ibaizabal River.

The concentrations measured in different coastal regions of the world have been compiled in Table 1. Although the concentration of the trace metals varied from area to area Fe, Cu, Mn and Zn are predominant in almost all of them.

In general Fe and Cu show the widest concentration ranges, with values between 0.27-41868 $mg{\cdot}kg^{-1}$ (Fe) and belong 1.0 and 2890 $mg{\cdot}kg^{-1}$ (Cu). The presence of As, Cd, Cr and Ni also vary from area to area, but in a lower extent. On the contrary, the concentrations found for Zn and Mn do no change significantly from zone to zone, which indicate that their presence in transitional areas may be usual (regardless their origin).

Pichavaran estuary in India presents the highest concentrations for three of the trace elements considered (Cd, Cr and Ni). Restronguent estuary in England and Canal de Santo Padre in Spain are in the second position, with significantly higher Cu and Zn, and As and Pb contents, respectively. Sediments from the Salt-Water estuary in Taiwan and Patras Harbour in Greece are the third on the list, with high contents of Fe and Mn.

Although the other aquatic environments present lower concentrations, some of them, as the estuary of Tagus in Portugal and Tamiraparani in India, have on its sediments considerable amounts of all the pollutants considered.

A deep examination of the pollution sources of the aquatic areas compiled in Table 1 may allow us to identify some general sources of trace elements to estuaries. All the areas suffer the pressures coming from cities or residential areas sited around their banks, which supposes a direct or the indirect discharge of urban and domestic wastewaters into the estuary.

Table 1. Concentrations (in mg·kg^{-1}) of several elements in sediments of different aquatic environments of the world

Estuaries	As	Cd	Cr	Cu	Fe	Mn	Ni	Pb	Zn	Ref.
Krka (Croatia)	-	-	10.67	16	-	232	5.7	11	9.7	(Prohic and Kniewald, 1987)
Ebro River (Spain)	-	0.23-1.5	-	2.2-37	-	-	-	2.8-70	21-198	(Ramos et al., 1999)
St Lucie (U.S.A.)	-	4.9-23	5.2-91	3.1-72	-	-	13-55	2.8-23	3.4-127	(Zhang et al., 2003)
Vigo (Spain)	-	-	33	26	23500	-	23	57	96	(Rubio et al., 2000)
Medway (U.K.)	14	-	76	42	30282	315	28	67	138	(Spencer, 2002)
Mersey (U.K.)	-	-	36	40	-	-	-	65	242	(Harland et al., 2000)
Hudson (U.S.A)	-	-	-	78	34000	-	-	81	182	(Feng et al., 1998)
Pearl (China)	-	-	34-135	6.2-100	-	-	11-54	16-93	55.1-268	(Carman et al., 2007)
Severn (U.K)	-	0.05-0.92	43-80	24-46	-	-	19-27	50-68	148-237	(Duquesne et al., 2006)
Tagus (Portugal)	-	1.7-5.9	-	28-89	-	-	-	65-199	-	(França et al., 2005)
Ulla (Spain)	-	-	191-438	96-135	30900-42100	495-521	69-91	57-58	134-190	(Prego and Cobelo-García, 2003)
Thames (England)	15	-	36	24	21711	328	21	63	115	(Attrill and Thomes, 1995)
MaraBasco (Mexico)		0.05-0.34	-	0.7-31	-	-	9-26	2-18	53-179	(Marmolejo-Rodriguez et al., 2007)
Pontevedra (Spain)	-	-	30-280	6-227	-	-	3-21	37-144	38-243	(Fernández et al., 2008)
Ennore Creek (India)	-	0.51	49	22	2023	226	18	32	96.7	(Jayaprakash et al., 2008)
Tamaki (New Zealand)		0.11-1.0	-	21-47	-	-	-	51-122	138-272	(Abrahim and Parker, 2008)
Pichavaram (India)	-	35	617	132	24998	801	252	144	106	(Ranjan et al., 2008)
Salt-Water (Taiwan)	-	1.4	131	1001	41868	639	103	128	1220	(Lin et al., 2011)
Nanliu (China)	18	0.1	45.3	68	-	-	-	34	57.4	(Xia et al., 2011)
Odiel (Spain)	330	14	-	2109	-	-	-	590	1154	(Vicente-Martorell et al., 2009)

Table 1. (Continued)

Estuaries	As	Cd	Cr	Cu	Fe	Mn	Ni	Pb	Zn	Ref.
Tinto (Spain)	339	8	-	1897	-	-	-	496	1115	(Vicente-Martorell et al., 2009)
Canal del Santo padre (Spain)	384	7	-	2215	-			630	1431	(Vicente-Martorell et al., 2009)
Cochin (India)	-	7	-	1346	44	4.04	70.6	27	578	(Harikumar and Nasir, 2010)
Minho (Portugal)	-	0.2-0.3	-	2.8-22	1.4-2.0	-	-	4.8-16	38-92	(Almeida et al., 2008)
Douro (Portugal)	-	0.1-0.3	-	1.0-229	0.27-3.0	-	-	0.25-192	6.2-457	(Almeida et al., 2008)
Tagus (Portugal)	-	0.9-11	-	9.7-214	3.1-5.0	-	-	11.8-350	88-1086	(Almeida et al., 2008)
Tamar (England)	-	0.4-0.5	-	129-161	30500-36500	348-390	25-28	139-165	246-282	(Singh and Turner, 2009)
Poxim (Brazil)	-	0.06-0.4	3-13	5.3-12	-	-	0.3-3	8.1-16.0	7.5-31	(de Andrade Passos et al., 2010)
Bohai Bay (China)		0.16-0.18	53-63	26-33	-	-	31-36	24-26	72-97	(Meng et al., 2008)
Gao-ping (Taiwan)	11	0.1	57	33	-	-	34	25	122	(Doong et al., 2008)
Patras Harbour (Greece)	8.8	0.5	202	82	35000	1015	110	49	120	(Papaefthymiou et al., 2010)
Thermaikos (Greece)	-	-	47	87	-	-	-	77	184	(Christophoridis et al., 2009)
Drin (Albania)	-	0.1-0.13	-	12-13	725-749	289-299	-	0.4-0.5	6.9-7.1	(Ianni et al., 2010)
Mahanadi (India)	-	1.4	3.1	4	141	182	8	7.9	20	(Sundaray et al., 2011)
Gironde (France)	19	0.5	78	24	-	-	32	47	168	(Larrose et al., 2010)
Tamiraparani (India)	-	11	4.23	40.2	16400	668	24	26	199	(Magesh et al., 2011)
Mississippi Gulf (USA)	13	0.6	83	-	22000	431	32	37	212	(Warren et al., 2012)
Nerbioi-Ibaizabal	0.6-219	0.05-17	5-134	13-769	5527-36715	67-936	3.4-936	21-458	41-2060	(Gredilla et al., 2012a)
Urdaibai (This study)	3.2-15	0.4-0.6	13-118	13-73	5238-8263	103-268	6.0-48	20-5.4	144-204	This study

Besides urban wastes, industrial activity is the most common source of metals, as in the case of the Ennore Creek in India, Sat-Water River in Taiwan or Tinto, Ulla and Odiel Rivers in Spain. Mining activities are also important pollution sources in coastal environments, as in the case of some estuaries sited in England (Restroguet or Gannel, for example) an in Spain (Tinto River).

Agricultural wastewaters are also important pollution sources for these areas. Activities related to agricultural works are the main pollution sources in areas such as Sado in Portugal or Nanliu River estuary in China.

Although it is still not so common, possible sources related to natural disasters start to be investigated. The trace element concentrations in surface estuarine sediments were analysed in the Mississippi Gulf Coast following Hurricane Katrina (Warren et al., 2012). Similar study was carried out in Pichavaram (India), with the aim of assessing heavy metal contamination after a tsunami (Ranjan et al., 2008).

As estuaries are largely used as communication links between different areas, they usually suffer from water-traffic pressures. This is the case of the Patras Harbour in Greece which has been though its history an important commercial and tourist link between Greece and Italy.

Although important measures have been adopted in order to minimise these inputs, there are still some areas of the world which dump untreated domestic and industrial wastes directly to the water. This is the case of Mahanadi basin in India and Poxim River in Brazil, among others.

Broadly speaking it can be said that the range of trace elements concentrations in sediments from the estuary of Urdaibai is not worrying in comparison with other estuaries of the world (see Table 1). The concentration of the analysed trace elements considered in this work is considerably lower, or at least comparable, than those found in the bibliography for other estuaries of the world.

The case of the estuary of the Nerbioi-Ibaizabal River is significantly different. Although the range of trace elements concentration in sediments of Nerbioi-Ibaizabal is still not worrying in comparison with other estuaries, the maximum values measured for Ni for example, are over those ones measured in the rest of the estuaries.

The highest concentrations found for Cd, Mn, Pb, Fe and Zn are only exceeded in Pichavaram, Patras harbour, Canal de Santo Padre, Salt-Water and in Restronguet respectively. For Cr, only other two estuaries presented higher concentrations than the Nerbioi-Ibaizabal estuary.

It should be underlined that although the simple comparison with other estuaries gives the possibility to have a general idea about the pollution level of the sediments under study, this comparison should be done with care due to the specific geochemical characteristics of each emplacement and to the differences in the methods used for sediment analysis.

ACKNOWLEDGMENTS

This work has been financially supported by the Basque Government through the ETORTEK BERRILUR 3 (IE09-242) project. A. Rodriguez-Iruretagoiena, J. Bustamante and A. Gredilla are grateful to the UPV/EHU for their pre-doctoral and post-doctoral fellowships.

REFERENCES

Abrahim, G. and Parker, R. (2008). Assessment of heavy metal enrichment factors and the degree of contamination in marine sediments from Tamaki Estuary, Auckland, New Zealand. *Environ. Monit. Assess., 136*, 227-238.

Almeida, C. M., Mucha, A. P., Bordalo, A. A. and Vasconcelos, M. T. (2008). Influence of a salt marsh plant (Halimione portulacoides) on the concentrations and potential mobility of metals in sediments. *Sci. Total Environ., 403*, 188-195.

Arias, R., Barona, A., Ibarra-Berastegi, G., Aranguiz, I. and Elías, A. (2008). Assessment of metal contamination in dregded sediments using fractionation and Self-Organizing Maps. *J. Hazard. Mater., 151*, 78-85.

Attrill, M. J. and Thomes, R. M. (1995). Heavy metal concentrations in sediment from the Thames Estuary, UK. *Mar. Poll. Bull., 30*, 742-744.

Audry, S., Blanc, G., Schafer, J. and Robert, S. (2007). Effect of estuarine sediment resuspension on early diagenesis, sulfide oxidation and dissolved molybdenum and uranium distribution in the Gironde estuary, France. *Chem. Geol., 238*, 149-167.

Bartolome, L., Tueros, I., Cortazar, E., Raposo, J. C., Sanz, J., Zuloaga, O., de, D.A., Etxebarria, N., Fernandez, L. A. and Madariaga, J. M. (2006). Distribution of trace organic contaminants and total mercury in sediments from the Bilbao and Urdaibai Estuaries (Bay of Biscay). *Mar. Poll. Bull., 52*, 1111-1117.

Belzunce, M. J., Solaun, O., Franco, J., Valencia, V. and Borja, A. (2001). Accumulation of Organic Matter, Heavy Metals and Organic Compounds in Surface Sediments along the Nervión Estuary (Northern Spain). *Mar. Poll. Bull., 42*, 1407-1411.

Carman, I., Xiang-Dong, L., Gan, Z., Onyx, W. and Yok-Sheung, L. (2007). Trace metal distribution in sediments of the Pearl River Estuary and the surrounding coastal area, South China. *Environ. Pollut., 147*, 311-323.

Cearreta, A., Irabien, M. J., Leorri, E., Yusta, I., Croudace, I. W. and Cundy, A. B. (2000). Recent Anthropogenic Impacts on the Bilbao Estuary, Northern Spain: Geochemical and Microfaunal Evidence. *Estuar. Coast. Shelf Sci., 50*, 571-592.

Cearreta, A., Irabien, M. J., Leorri, E., Yusta, I., Quintanilla, A. and Zabaleta, A. (2002). Environmental transformation of the Bilbao estuary, N. Spain: microfaunal and geochemical proxies in the recent sedimentary record. *Mar. Poll. Bull., 44*, 487-503.

Censi, P., Spoto, S. E., Saiano, F., Sprovieri, M., Mazzola, S., Nardone, G., Di Geronimo, S. I., Punturo, R. and Ottonello, D. (2006). Heavy metals in coastal water systems. A case study from the northwestern Gulf of Thailand. *Chemosphere, 64*, 1167-1176.

Christophoridis, C., Dedepsidis, D. and Fytianos, K. (2009). Occurrence and distribution of selected heavy metals in the surface sediments of Thermaikos Gulf, N. Greece. Assessment using pollution indicators. *J. Hazard. Mater., 168*, 1082-1091.

Cundy, A. B., Croudace, I. W., Cearreta, A. and Irabien, M. J. (2003). Reconstructing historical trends in metal input in heavily-disturbed, contaminated estuaries: studies from Bilbao, Southampton Water and Sicily. *Appl. Geochem., 18*, 311-325.

Dai, M., Martin, J.-M. and Cauwet, G. (1995). The significant role of colloids in the transport and transformation of organic carbon and associated trace metals (Cd, Cu and Ni) in the Rhone delta (France). *Mar. Chem., 51*, 159-175.

de Andrade Passos, E., Alves, J. C., dos Santos, I. S., Alves, J. P. H., Garcia, C. A. B. and Spinola Costa, A. C. (2010). Assessment of trace metals contamination in estuarine sediments using a sequential extraction technique and principal component analysis. *Microchem. J., 96*, 50-57.

Demirak, A., Yilmaz, F., Levent Tuna, A. and Ozdemir, N. (2006). Heavy metals in water, sediment and tissues of Leuciscus cephalus from a stream in southwestern Turkey. *Chemosphere, 63*, 1451-1458.

Doong, R. A., Lee, S. H., Lee, C. C., Sun, Y. C. and Wu, S. C. (2008). Characterization and composition of heavy metals and persistent organic pollutants in water and estuarine sediments from Gao-ping River, Taiwan. *Mar. Poll. Bull., 57*, 846-857.

Duquesne, S., Newton, L. C., Giusti, L., Marriott, S. B., Stärk, H.-J. and Bird, D. J. (2006). Evidence for declining levels of heavy-metals in the Severn Estuary and Bristol Channel, U.K. and their spatial distribution in sediments. *Environ. Pollut., 143*, 187-196.

El-Moselhy, K. M. and Gabal, M. N. (2004). Trace metals in water, sediments and marine organisms from the northern part of the Gulf of Suez, Red Sea. *J. Mar. Syst., 46*, 39-46.

Fdez-Ortiz de Vallejuelo, S., Arana, G., de Diego, A. and Madariaga, J. M. (2010). Risk assessment of trace elements in sediments: The case of the estuary of the Nerbioi-Ibaizabal River (Basque Country). *J. Hazard. Mater., 181*, 565-573.

Fdez-Ortiz de Vallejuelo, S., Arana, G., de Diego, A. and Madariaga, J. M. (2011). Pattern recognition and classification of sediments according to their metal content using chemometric tools. A case study: The estuary of Nerbioi-Ibaizabal River (Bilbao, Basque Country). *Chemosphere, 85*, 1347-1352.

Fdez-Ortiz de Vallejuelo, S., Barrena, A., Arana, G., de Diego, A. and Madariaga, J. M. (2009). Ultrasound energy focused in a glass probe: An approach to the simultaneous and fast extraction of trace elements from sediments. *Talanta, 80*, 434-439.

Feng, H., Kirk Cochran, J., Lwiza, H., Brownawell, B. J. and Hirschberg, D. J. (1998). Distribution of heavy metal and PCB contaminants in the sediments of an urban estuary: The Hudson River. *Mar. Environ. Res., 45*, 69-88.

Fernández, N., Bellas, J., Lorenzo, J. and Beiras, R. (2008). Complementary Approaches to Assess the Environmental Quality of Estuarine Sediments. *Water Air Soil Poll., 189*, 163-177.

Franca, S., Vinagre, C., Caçador, I. and Cabral, H. N. (2005). Heavy metal concentrations in sediment, benthic invertebrates and fish in three salt marsh areas subjected to different pollution loads in the Tagus Estuary (Portugal). *Mar. Poll. Bull., 50*, 998-1003.

Gredilla, A., Fdez-Ortiz de Vallejuelo, Arana, G., de Diego, A. and Madariaga, J. M. (2012a). Long term seasonal monitoring of trace element pollution in sediments from the estuary of teh Nerbioi-Ibaizabal River. *Submitted to Journal of Soil and Sediment.*

Gredilla, A., Fdez-Ortiz de Vallejuelo, Arana, G., de Diego, A. and Madariaga, J. M. (2012b). A new methodology to investigate chemical pollution in aquatic systems: from the local scale to a wide world perspective. A case study: the estuary of the Nerbioi-Ibaizabal River (Bilbao, Basque Country). *Submitted to Chemosphere.*

Hakanson, L. (1980). An ecological risk index for aquatic pollution control. A sedimentological approach. *Water Res., 14*, 975-1001.

Harikumar, P. S. and Nasir, U. P. (2010). Ecotoxicological impact assessment of heavy metals in core sediments of a tropical estuary. *Ecotoxicol. Environ. Saf., 73*, 1742-1747.

Harland, B. J., Taylor, D. and Wither, A. (2000). The distribution of mercury and other trace metals in the sediments of the Mersey Estuary over 25 years 1974-1998. *Sci. Total Environ., 253*, 45-62.

Ianni, C., Bignasca, A., Magi, E. and Rivaro, P. (2010). Metal bioavailability in marine sediments measured by chemical extraction and enzymatic mobilization. *Microchem. J., 96*, 308-316.

Irabien, M. J. and Velasco, F. (1999). Heavy metals in Oka river sediments (Urdaibai National Biosphere Reserve, northern Spain): lithogenic and anthropogenic effects. *Environ. Geol., 37*, 54-63.

Jayaprakash, M., Jonathan, M. P., Srinivasalu, S., Muthuraj, S., Ram-Mohan, V. and Rajeshwara-Rao, N. (2008). Acid-leachable trace metals in sediments from an industrialized region (Ennore Creek) of Chennai City, SE coast of India: An approach towards regular monitoring. *Estuar. Coast. Shelf Sci., 76*, 692-703.

Larrose, A., Coynel, A., Schäfer, J., Blanc, G., Massé, L. and Maneux, E. (2010). Assessing the current state of the Gironde Estuary by mapping priority contaminant distribution and risk potential in surface sediment. *Appl. Geochem., 25*, 1912-1923.

Leorri, E., Cearreta, A., Irabien, M. J. and Yusta, I. (2008). Geochemical and microfaunal proxies to assess environmental quality conditions during the recovery process of a heavily polluted estuary: The Bilbao estuary case (N. Spain). *Sci. Total Environ., 396*, 12-27.

Lin, C. E., Chen, C. T., Kao, C. M., Hong, A. and Wu, C. Y. (2011). Development of the sediment and water quality management strategies for the Salt-water River, Taiwan. *Mar. Poll. Bull., 63*, 528-534.

Magesh, N. S., Chandrasekar, N. and Vetha Roy, D. (2011). Spatial analysis of trace element contamination in sediments of Tamiraparani estuary, southeast coast of India. *Estuar. Coast. Shelf Sci., 92*, 618-628.

Marmolejo-Rodriguez, A. J., Prego, R., Meyer-Willerer, A., Shumilin, E. and Cobelo-Garcia, A. (2007). Total and labile metals in surface sediments of the tropical river-estuary system of Marabasco (Pacific coast of Mexico): Influence of an iron mine. *Mar. Poll. Bull., 55*, 459-468.

Mendiguchía Martínez, C. (2005). Utilización de ultratrazas de metales pesados como trazadores de los efectos antropogénicos producidos en ecosistemas acuáticos. Doctoral Thesis.

Meng, W., Qin, Y., Zheng, B. and Zhang, L. (2008). Heavy metal pollution in Tianjin Bohai Bay. *J. Environ. Sci., 20*, 814-819.

Müller, G. (1979). Schwermetalle in den sediment des Rheins, Veranderungem Seit 1971. *Umschau, 79*, 778-783.

Muñoz Barbosa, A., Gutiérrez-Galindo, E. A., Daesslé, L. W., Orozco-Borbón, M. V. and Segovia-Zavala, J. A. (2010). Relationship between metal enrichments and a biological adverse effects index in sediments from Todos Santos Bay, northwest coast of Baja California, México. *Mar. Poll. Bull., 64*, 405-409.

Orive, E., Franco, J. and Ruiz, A. (1995). I*mportancia del fitoplancton en estuarios meso-macromareales someros: el ejemplo del estuario de Urdaibai*. Angulo E. (Ed.), Urdaibai: investigación basica y aplicada, Gobierno Vasco, pp.57-74.

Orive, E., Iriarte, A., De Madariaga, I. and Revilla, M. (1998). Phytoplankton blooms in the Urdaibai estuary during summer: Physico-chemical conditions and taxa involved. *Oceanol. Acta, 21*, 293-305.

Papaefthymiou, H., Papatheodorou, G., Christodoulou, D., Geraga, M., Moustakli, A. and Kapolos, J. (2010). Elemental concentrations in sediments of the Patras Harbour, Greece, using INAA, ICP-MS and AAS. *Microchem. J., 96*, 269-276.

Prego, R. and Cobelo-García, A. (2003). Twentieth century overview of heavy metals in the Galician Rias (NW Iberian Peninsula). *Environ. Pollut., 121*, 425-452.

Prohic, E. and Kniewald, G. (1987). Heavy metal distribution in recent sediments of the Krka River Estuary -- an example of sequential extraction analysis. *Mar. Chem., 22*, 279-297.

Puy-Azurmendi, E., Ortiz-Zarragoitia, M., Kuster, M., Martínez, E., Guillamón, M., Domínguez, C., Serrano, T., Barbero, M. C., Alda, M.L., Bayona, J. M., Barceló, D. and Cajaraville, M. P. (2006). An integrated study of endocrine disruptors in sediments and reproduction-related parameters in bivalve molluscs from the Biosphere´s Reserve of Urdaibai (Bay of Biscay). *Mar. Environ. Res., 69, Supplement 1*, S63-S66.

Rainbow, P. S., Kriefman, S., Smith, B. D. and Luoma, S. N. (2011). Have the bioavailabilities of trace metals to a suite of biomonitors changed over three decades in SW England estuaries historically affected by mining? *Sci. Total Environ., 409*, 1589-1602.

Ramos, L., Fernández, M. A., González, M. J. and Hernández, L. M. (1999). Heavy Metal Pollution in Water, Sediments, and Earthworms from the Ebro River, Spain. *B. Environ. Contam. Tox., 63*, 305-311.

Ranjan, R. K., Ramanathan, A., Singh, G. and Chidambaram, S. (2008). Assessment of metal enrichments in tsunamigenic sediments of Pichavaram mangroves, southeast coast of India. *Environ. Monit. Assess., 147*, 389-411.

Rodríguez, J. G., Tueros, I., Borja, A., Belzunce, M. J., Franco, J., Solaun, O., Valencia, V. and Zuazo, A. (2006). Maximum likelihood mixture estimation to determine metal background values in estuarine and coastal sediments within the European Water Framework Directive. *Sci. Total Environ., 370*, 278-293.

Rubio, B., Nombela, M. A. and Vilas, F. (2000). Geochemistry of Major and Trace Elements in Sediments of the Ria de Vigo (NW Spain): an Assessment of Metal Pollution. *Mar. Poll. Bull., 40*, 968-980.

Scancar, J., Milacic, R., Strazar, M. and Burica, O. (2000). Total metal concentrations and partitioning of Cd, Cr, Cu, Fe, Ni and Zn in sewage sludge. *Sci. Total Environ., 250*, 9-19.

Seebold, I., Labarta, C. and Amigó, J. M. (1982). *Heavy metals in the sediments of the Bilbao estuary.* Oxford, Pergamon Press.

Singh, A. and Turner, A. (2009). Surfactant-induced mobilisation of trace metals from estuarine sediment: Implications for contaminant bioaccessibility and remediation. *Environ. Pollut., 157*, 646-653.

Spencer, K. L. (2002). Spatial variability of metals in the inter-tidal sediments of the Medway Estuary, Kent, UK. *Mar. Poll. Bull., 44*, 933-944.

Spencer, K. L., Cundy, A. B. and Croudace, I. W. (2003). Heavy metal distribution and early-diagenesis in salt marsh sediments from the Medway Estuary, Kent, UK. *Estuar. Coast. Shelf Sci., 57*, 43-54.

Sundaray, S. K., Nayak, B. B., Lin, S. and Bhatta, D. (2011). Geochemical speciation and risk assessment of heavy metals in the river estuarine sediments—A case study: Mahanadi basin, India. *J. Hazard. Mater., 186*, 1837-1846.

Szava-Kovats, R. C. (2008). Grain-size normalization as a tool to assess contamination in marine sediments: Is the <63 μm fraction fine enough? *Mar. Poll. Bull., 56*, 629-632.

Telesh, I. V. and Khlebovich, V. V. (2010). Principal processes within the estuarine salinity gradient: A review. *Mar. Poll. Bull., 61*, 149-155.

Tueros, I., Borja, A., Larreta, J., Rodríguez, J. G., Valencia, V. and Millán, E. (2009). Integrating long-term water and sediment pollution data, in assessing chemical status within the European Water Framework Directive. *Mar. Poll. Bull., 58*, 1389-1400.

Vicente-Martorell, J. J., Galindo-Riaño, M. D., García-Vargas, M. and Granado-Castro, M. D. (2009). Bioavailability of heavy metals monitoring water, sediments and fish species from a polluted estuary. *J. Hazard. Mater., 162*, 823-836.

Warren, C., Duzgoren-Aydin, N. S., Weston, J. and Willett, K. L. (2012). Trace element concentrations in surface estuarine and marine sediments along the Mississippi Gulf Coast following Hurricane Katrina. *Environ. Monit. Assess., 184*, 1107-1119.

WFD (2000). *Directive of the European Parliament and of the Council* 2000/60/EC Establishing a Framework for Community Action in the Field of Water Policy, European Comission Environment, Louxenburg.

Xia, P., Meng, X., Yin, P., Cao, Z. and Wang, X. (2011). Eighty-year sedimentary record of heavy metal inputs in the intertidal sediments from the Nanliu River estuary, Beibu Gulf of South China Sea. *Environ. Pollut., 159*, 92-99.

Zhang, M., He, Z., Stoffella, P., Calvert, D., Yang, X., and Sime, P. (2003). Concentrations and solubility of heavy metals in muck sediments from the St. Lucie Estuary, U.S.A. *Environ. Geol., 44*, 1-7.

In: Trace Elements ISBN: 978-1-62081-376-8
Editors: Diego A. De Leon and Paloma Raquel Aragon

Chapter 6

DISTINGUISHING BETWEEN NATURAL AND ANTHROPOGENIC SOURCES OF TRACE ELEMENTS IN THE SEDIMENTS USING THE METHODS OF GEOCHEMICAL NORMALIZATION AND STATISTICAL ANALYSIS

Sanja M. Sakan* and Dragana S. Đorđević
ICTM, Department of Chemistry, University of Belgrade, Belgrade, Serbia

ABSTRACT

The contamination of sediments by potentially harmful trace elements is of major concern, because of their toxicity, persistence and bioaccumulative nature. In many industrial and urban areas, sediments are the largest repository for, and potential source of these elements.

Assessing trace element contamination of sediment is complicated since these elements are a ubiquitous, naturally occurring component of sediment, their concentrations in un-contaminated sediment can vary by orders of magnitude over relatively small spatial scales, and naturally occurring and anthropogenically introduced metals tend to accumulate in the same areas. Due to their particle reactivity, trace elements tend to accumulate in sediments and may persist in the environment long after their primary source has been removed. Element adsorption increases with decreasing grain size of the sediment. Thus, the metal concentrations significantly enriched in fine-grained sediment rich in clay minerals. Because of that, in geochemical and ecochemical studies, the normalization of elemental content using an immobile element is common practice for correcting grain-size effects and dilution by sedimentary phases such as carbonates and silica.

The aims of the presented study were: (1) to assess the geochemistry of trace elements (Cr, Zn, Cd, Ni, Pb, and Cu) in canal sediments (Danube alluvial formation) and Tisa River sediments and its tributaries, after normalization of the elements data to a conservative element; (2) to discriminate natural and anthropogenic contributions; and (3)

* Email: ssakan@chem.bg.ac.rs.

to identify possible sources of pollution. As the element for sediment normalization was used Al, since it represents aluminosilicates, the main group of minerals generally found in the fine sediment fractions. Flame atomic absorption spectrometry (FAAS) was used to determine the levels of the investigated elements. Multivariate statistical methods have been applied in this research to identify similarities in the sampling stations and to assess the relationship between the elements.

The results from this study show that investigated sediments are affected by combined trace elements contamination. Also, there was shown a significance of applying geochemical normalization and statistical analysis for assessing trace element contamination and for distinguishing between natural and anthropogenic sources of trace elements in the sediments.

1. Introduction

The literature includes a range of explanations for definition of "trace element", depending on the scientific approach. In analytical chemistry, a trace element is an element in a sample that has an average concentration of less than 100 parts per million measured in atomic count, or less than 100 micrograms per gram. In geochemistry, a trace element is a chemical element whose concentration is less than 1000 ppm or 0.1% of a rock's composition, whereas in biochemistry, a trace element is an element that occurs in the body in mg kg^{-1} of body weight or less. Currently, the term "trace element" is increasingly preferred in environmental literature to the meaningless term "heavy metal", which was widely used in the recent past (and is still used) to denominate a group name for metals and semimetals (i.e. metalloids) that have been associated with contamination and potential toxicity or ecotoxicity. [1].

Investigations concerning the trace elements present in sediments have increased in recent years [2-10]. Previously, sediments have been considered as reservoirs into which chemical species are deposited. They are now viewed as an active aquatic compartment that performs a fundamental role in redistribution of these species to aquatic biota [11]. Potentially harmful trace elements and contamination with these elements have been the subject of much attention because of their peculiar pollutant characteristics: (1) they do not decay with time, unlike many organics and radionuclides; (2) they can be necessary or beneficial to plants at certain levels, but can also be toxic when exceeding specific thresholds; (3) they are always present at a background level of non-anthropic origin, their input in soils being related to weathering of parent rocks and pedogenesis, and (4) they often occur as cations, which strongly interact with soil matrix; consequently, heavy metals in soils, even at high concentrations, may be present in inert and not harmful forms, but they can become mobile as a result of changing environmental conditions (land use, agricultural input, climatic change) or by saturation beyond the buffering capacity of the soil. This situation is referred to as a "chemical time bomb" [12] and sediments are secondary sources of contaminants in aquatic system.

The trace metal variability of sediments may be natural or influenced to some degree by anthropogenic sources. Trace elements are naturally occurring constituents on the Earth and vary in concentration in air, rock, soil, sediments and biota across geographic regions [13]. However, since the advent of man, anthropogenic inputs have overwhelmed the natural trace element cycling and have led to increased concentrations and an accumulation of trace elements in the environment as well as strong interference in the transport fluxes and rates between biogeochemical compartments in both terrestrial and aquatic ecosystems [14]. Therefore, trace element contamination (i.e. the presence of a trace element where it should not be or at concentrations above the natural background) is unquestionably present everywhere on the Earth [1].

The ability of sediments to adsorb organic and inorganic contaminants makes sediment analysis a valuable tool with which to assess and monitor water and sediment quality and track contaminant migration in the environment. This ability is dependent largely on the grain-size distribution of sediments; fine-grained sediment adsorbs contaminants more readily than coarse-grained sediment [15]. Thus, the metal concentrations significantly enriched in fine-grained sediment rich in clay minerals. Because of that, in geochemical and ecochemical studies, the normalization of elemental content using an immobile element is common practice for correcting grain-size effects and dilution by sedimentary phases such as carbonates and silica. In order to establish background contaminant levels and identification potential outliers, 'normalization' is required to mitigate differences in grain-size distribution among sediment samples. Normalization is usually performed for a suite of samples by determining the relationship between an element or compound of interest and either a 'conservative' element - one presumed to have no anthropogenic enrichment or the amount of fine-grained material in each sample. This simple way of data processing may also reveal possible sediment enrichment in trace metals.

By making trace metal data from different sediment types more directly comparable, normalization can provide information on contaminant sources. Heavy metals, like cadmium, chromium, copper, nickel, lead, and zinc are most important among the frequently observed contaminants in sediments. Al has been the most widely used element for normalization, because it is a major constituent of fine-grained aluminosilicates with which the bulk of trace metals are associated [16].

The aims of the presented study were: (1) to assess the geochemistry of trace elements (Cr, Zn, Cd, Ni, Pb, and Cu) in canal sediments (Danube alluvial formation) and Tisa River sediments and its tributaries, after normalization of the elements data to a conservative element; (2) to discriminate natural and anthropogenic contributions; and (3) to identify possible sources of pollution. As the element for sediment normalization was used Al, since it represents aluminosilicates, the main group of minerals generally found in the fine sediment fractions. Flame atomic absorption spectrometry (FAAS) was used to determine the levels of the investigated elements. Multivariate statistical methods have been applied in this research to identify similarities in the sampling stations and to assess the relationship between the elements.

2. EXPERIMENTAL

2.1. Study Area, Sampling and Sample Preservation

The studied area represented sediments that were taken at two locations. One part of the samples is sediments of the River Tisa and its tributaries, and the second part of the samples is alluvial sediments from the Danube, which were taken in the industrial zone of the Pančevo.

Numerous industrial installations located on the banks of the Danube or its tributaries have contributed to important anthropogenic contamination. Figure 1 shows the locations of the sampling sites.

The investigated alluvial sediments were taken from the canal located in the industrial zone south-east of the City of Pančevo, about 15 km east of Belgrade (Figure 1).

The most important economic industries in the Pančevo are crude oil processing, HIP Petrochemical Complex, fertilizers (Nitrogen Plant), the textile and fashion industry and agriculture. The samples of alluvial sediments were taken during the 2001, from 14 drill-holes (marked as S1-S14) at different depths, up to 5m.

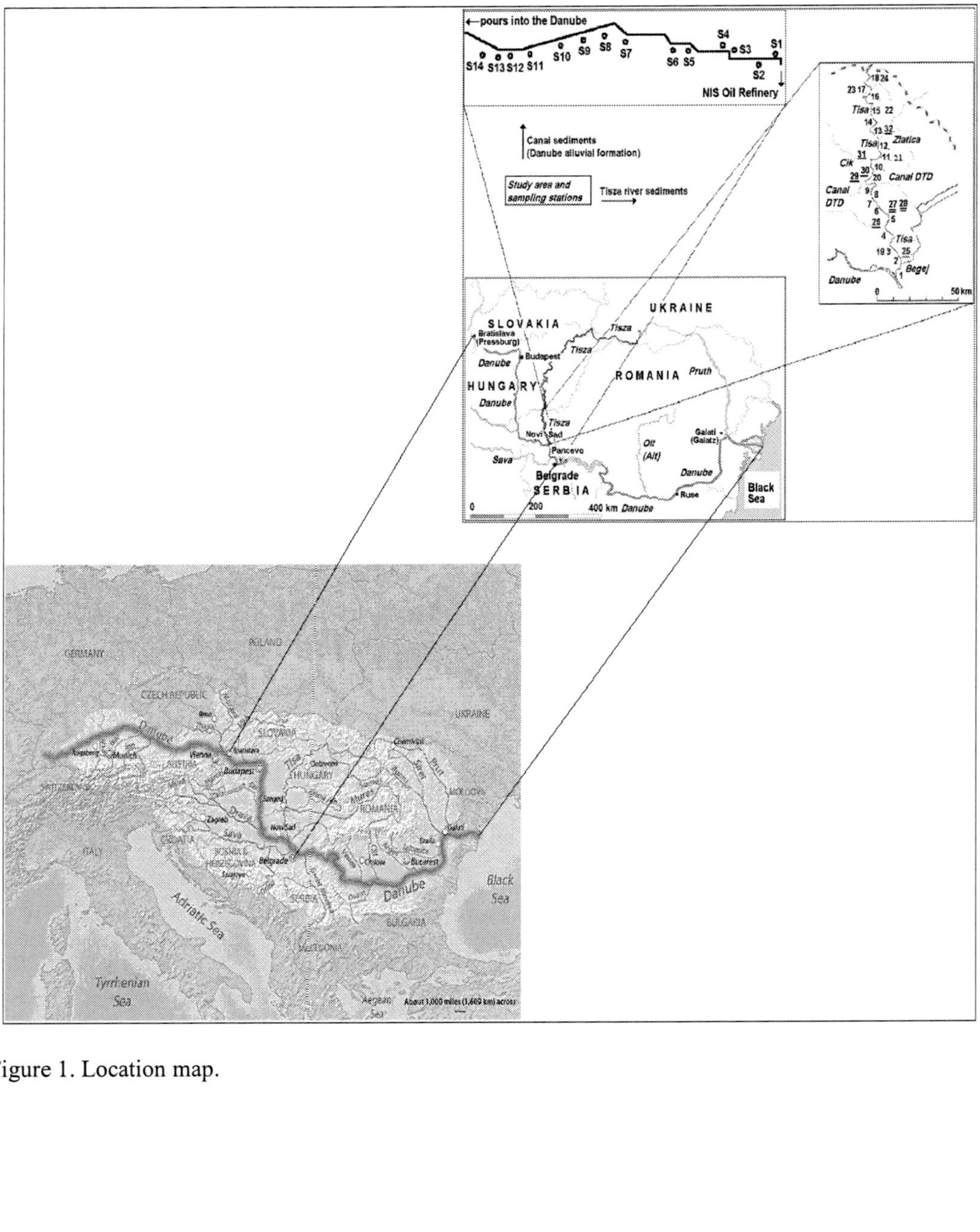

Figure 1. Location map.

The Tisa River region is one of the largest natural riverside systems in Central and South-Eastern Europe. It flows roughly along the Romanian border and enters Hungary at Tiszabecs; downstream, it marks the Slovak–Hungarian border, passes through Hungary and flows into the Danube in northern Serbia (Vojvodina). Numerous industrial accidents have contaminated the river with hazardous substances during the last several decades.

In addition, in the Tisa Basin are the Mantra Mountains of eastern Hungary, which comprise magmatic exceptions to Hungary's sedimentary-dominated geology.

Mining in the Mantra Mountains has focused on the Cu/Au and the Pb/Zn deposits [17]. The Serbian part of the Tisa watershed was selected for the present investigations. Sediment samples were collected during the 2001, from 32 locations: 24 sampling sites of the River Tisa and 8 sampling sites of tributaries and pools. After sampling, the samples were packed in pouches and frozen in order to prevent changes in chemical composition of the sediment.

2.2. Determination of the Metal Contents

2.2.1. Extraction Method

For extraction of elements, in this paper was applied method of sequential extraction. This method is described in detail in the papers [3,4,6] and the results of sequential extraction were discussed in the previous papers by the author.

In this paper were used the results of total extractable content elements. For the extraction of elements were used reagents: (1) the first step (exchangeable fraction, F1): 10 g sediment sample was extracted with 1 M $CH_3COO(NH_4)$; (2) the second step (represents metals bound to carbonates and easily reducible phases, F_2): extractant 0.6 M HCl and 0.1 M $NH_2OH{\cdot}HCl$; (3) the third step (metals bound to moderately reducible phases, F_3): a mixture of 0.2 M $H_2C_2O_4/(NH_4)_2C_2O_4$ was used as the extractant; (4) the fourth step (metals bound to organic matter and sulfides, F4): extractant 30% H_2O_2 adjusted to pH 2.0 with HNO_3 and (5) the fifth step (residual fraction, F5): extractant 6 M HCl.

The total amounts of elements in the sediments examined in this study are defined as the total extractable content and represent the sum of the elements released in all the five binding fractions.

2.2.2. Metal Determination

Flame atomic absorption spectrometry (FAAS) was used to determine the levels of the trace elements (SpectraAA55 Varian spectrophotometer, equipped with a hydride vapour system). External standard solutions were prepared from 1000 mg L^{-1} stock metal solutions. For minimized interferences, a multi-element standard stock solution was prepared in which the ratios of the metals in the multiple element calibration standards were analogous to their ratios in the samples. In order to simulate the composition of the investigated samples, the multi-element standards also contained elements which were not determined. Possible interference effects were minimized in this manner. As a quality control, duplicate analyses were performed on all samples and the precision was controlled. The relative standard deviations of the means of duplicate measurement were less than 10%.

The values of limit of detection (*DL*) for each determined element, expressed in μg m L^{-1}, were: Cr (0.03), Zn (0.005), Pb (0.05), Cd (0.01), Ni (0.05), and Cu (0.02). All concentrations values below the detection limit were replaced by half of the *DL*, as suggested by Reimann and Filmozer [18] and Relić *et al.* [19].

The moisture content of each sample was determined by drying a separate 1 g sample in an oven (105 ± 2 °C) to constant weight. From this, a correction to dry mass was obtained, which was applied to all reported element concentrations.

2.3. Normalization to Al

The main assumption for the application of a geochemical normalization to a conservative element is the existence of a linear relationship between the normalizer and other metals.

Such a relationship suggests that, in the natural sediments of the area, the concentration of the metal will change proportionally to the concentration of the normalizer.

After the selection of elements for normalization, then it is possible to produce a scatter plot of the metal and normalizer concentrations for the natural sediments of the area and then test the rations metal/normalizer at other (possibly contaminated) areas.

For the production of such a plot it is necessary to remove outlier values and to delineate a confidence band of 95% of the regression line of the metal on the normalizer.

Then, the data-points the possibly contaminated areas have to be projected on the diagram. All points which are found inside the 95% confidence band can be characterized as natural sediments, while all pointes above this area should be characterized as contaminated sediments [16].

In this paper, normalization to Al was performed to identify the anthropogenic inputs of elements. For this purpose the total element contents were compared to the total content of Al, Li, Ti, and Si.

2.4. Statistical Analysis

PCA and derivative methods have been widely used in geochemical applications to identify pollution sources and to apportion natural vs. anthropic contribution. [12].

Trace element studies are usually carried out using univariate procedures (element to element, or sample to sample), but multivariate methods such as principal component analysis (PCA) can provide further interpretation of results.

PCA is a data reduction procedure whose main goal is to provide an easy visualization of the relationships existing amongst the variables determined in large or complex data sets [11]. PCA was performed using SPSS version 11.5 for Windows.

Table 1. Content of elements in investigated Tisa river sediment

Sample		Cu	Cr	Zn	Pb	Cd	Ni	Al	Ti	Li	Si
1	0 km[a]	73.28	13.26	330	55.24	4.62	28.49	34281	433	26.00	29649
2	10 km	34.06	6.99	181	30.91	2.68	23.74	19179	501	14.20	20553
3	21 km	86.42	17.34	386	74.07	4.76	36.94	40578	579	29.00	26261
4	30 km	94.57	15.59	428	84.92	4.97	39.40	43394	673	35.20	28244
5	40 km	93.93	22.10	460	72.09	3.89	39.98	47530	645	34.10	27754
6	50 km	88.14	16.42	428	54.37	3.18	35.97	51266	570	36.70	33995
7	60 km	120	21.88	478	102	4.85	35.17	43297	769	32.10	28723
8	64 km	87.16	22.21	420	62.75	4.21	35.96	39691	634	29.00	20485
9	70 km	80.1	15.87	355	64.10	3.43	24.24	39284	933	26.70	22555
10	80 km	56.47	15.83	257	57.63	2.51	23.04	34746	644	26.20	18801
11	90 km	162	20.92	567	123	4.72	28.88	51262	1407	41.40	27827
12	100 km	71.29	14.88	318	69.26	2.83	27.71	29353	476	24.50	19598
13	110 km	110	20.54	394	74.22	2.91	32.30	20730	421	17.20	21904
14	120 km	102	20.43	383	59.17	4.00	25.49	20138	547	15.40	16293
15	130 km	132	22.57	404	75.33	3.39	33.23	30435	782	22.80	14476
16	140 km	87.83	18.85	334	57.30	2.44	25.67	31472	881	22.90	23975
17	150 km	94.60	19.56	406	86.11	4.37	23.87	31468	719	22.80	21578
18	158 km	72.05	20.28	275	68.23	3.34	21.62	28500	710	19.60	26096
19	21[b] km	109	24.44	399	72.36	2.71	41.20	25321	631	18.90	17473
20	70 km	67.93	16.93	282	44.62	2.31	24.01	24166	432	18.70	12084
21	90 km	54.92	13.85	250	38.04	2.25	17.28	28691	517	22.80	30367
22	130 km	58.70	14.40	245.	51.69	2.81	17.76	20919	614	16.90	21492
23	150 km	59.45	15.25	285	44.98	2.41	22.29	25889	948	19.70	23593
24	158 km	73.95	16.71	286	74.84	3.44	19.77	22072	535	16.80	18896
25	Begej[c]	45.80	11.82	135	21.08	2.04	48.85	30391	1228	44.40	19337
26	Jegrička	41.02	10.93	156	16.64	2.48	40.24	28598	1935	41.50	27647
27	Čikoš p.	33.44	9.84	53.94	15.61	3.12	19.95	16341	1428	17.00	23390
28	Okanj p.	32.32	7.30	99.40	11.17	2.36	30.46	31157	1762	40.70	29802
29	Canal B.	47.79	12.57	164	24.90	2.18	54.58	45205	1716	55.10	5631
30	Mrtva T.	40.90	11.75	138	21.47	2.49	45.63	35800	2587	42.90	38650
31	Čik	31.61	11.46	92.62	16.58	3.20	29.04	25011	2654	39.60	25929
32	Zlatica	54.37	13.77	107	22.07	3.03	47.12	31259	164	47.60	6465

Sampling sites are shown as numbers in respect to the distance from the Tisa mouth, as follows [a] Tisa, surface river sediment, and [b] Tisa buried river sediment; [c] Tisa tributaries and pools.

Table 2. Content of elements in investigated Danube alluvial sediment, Pančevo

Sample		Cu	Cr	Zn	Pb	Cd	Ni	Al	Ti	Li	Si
1	S1/1.0[a]	26.27	11.77	72.62	54.40	0.94	29.57	22199	255	23.00	18610
2	S1/2.4	33.52	12.77	77.51	42.56	0.62	35.72	30362	435	30.20	21300
3	S1/3.2	87.10	10.42	69.56	15.56	1.19	31.64	18066	203	26.40	16689
4	S2/1.0	25.99	14.45	65.55	18.69	1.21	29.38	21724	348	31.00	18535
5	S2/2.5	11.74	8.88	29.40	10.00	1.01	14.85	9666	530	18.60	16310
6	S2/5.0	10.77	8.62	31.45	11.97	0.95	12.48	8268	395	17.10	15802
7	S3/1.2	3.75	6.40	62.42	13.02	0.98	13.17	7781	468	13.10	13477
8	S3/2.2	19.69	11.60	56.38	17.51	1.07	23.14	15232	266	20.60	17678
9	S3/2.9	8.53	6.96	33.40	13.34	0.83	13.44	16167	940	22.90	31133
10	S4/1.0	3.72	5.52	163.60	13.15	0.81	16.17	3281	220	6.78	6052
11	S4/1.5	20.62	10.21	62.65	17.87	0.86	23.06	19140	450	21.10	16203
12	S4/2.4	6.39	5.96	23.53	17.86	0.86	12.40	3987	266	8.91	6814
13	S5/0.9	2.09	8.24	25.60	5.42	0.95	19.53	2956	265	5.81	5196
14	S6/0.9	2.83	7.83	39.28	6.06	1.33	16.69	4880	368	10.90	8408
15	S7/1.0	26.44	16.29	78.23	6.39	1.25	35.09	31168	734	13.60	24391
16	S7/1.5	3.51	9.87	27.34	3.25	0.89	16.27	5646	356	16.70	10087
17	S8/1.0	1.72	10.34	27.90	2.99	0.91	16.30	8651	447	8.48	9438
18	S9/0.9	1.74	7.49	25.21	3.14	0.93	15.59	10128	738	8.20	6905
19	S10/0.5	5.99	5.62	24.93	8.83	0.60	6.67	3022	178	6.24	4052
20	S10/0.8	23.48	6.71	41.69	6.23	1.47	8.36	13252	314	21.20	17537
21	S10/1.3	3.06	11.44	38.58	5.65	1.19	17.55	5333	374	9.27	7327
22	S11/0.0	21.55	18.70	84.30	8.11	1.51	21.68	24590	600	19.20	18512
23	S11/1.5	20.17	22.74	69.02	4.13	1.19	27.38	23225	791	24.00	12913
24	S12/0.3	32.83	26.71	111.37	8.48	1.88	24.88	36453	629	25.50	29271
25	S12/1.5	10.27	15.47	39.32	5.61	1.06	17.39	9990	454	32.30	7382
26	S12/2.3	28.23	23.12	76.76	6.08	1.18	39.50	25250	245	29.80	8876
27	S13/0.5	14.89	14.35	52.28	5.65	1.03	18.56	16766	514	18.70	10648
28	S13/1.4	6.16	10.37	29.83	4.99	1.05	13.88	7467	643	15.60	17797
29	S13/2.7	26.18	21.26	78.02	10.27	1.17	30.88	22797	691	29.80	15155
30	S14/0.7	12.11	15.57	45.55	5.35	1.12	19.61	11822	341	20.20	6456
31	S14/2.2	18.14	21.96	65.82	8.56	1.12	27.07	14982	645	26.20	9497

[a] S-number of drill hole and depth of sampling (m).

3. Results and Discussion

3.1. Analysis of the Results Obtained by Extraction

In Table 1 and Table 2 are presented results of total element content for investigated elements. From the data of Tables it is obvious that on some localities greater content of the extracted elements compared to the element content on other localities. In order to assess the existence of anthropogenic sources of these elements, method of geochemical normalization, statistical method and comparison with background values were applied to the obtained results.

3.2. The Results of the Geochemical Normalization

The suitability of Li, Ti, and Si against Al as a geochemical normalizer in the study areas was tested (for Tisa river and Pančevo alluvial sediments). Since that, a correlation matrix was calculated for trace element (Cu, Cr, Zn, Pb, Cd and Ni) and potential elements for normalization. Table 3 summarizes values of Pearson correlations coefficients (r) and significance level (p) for investigated elements in sediments from Pančevo (Danube alluvial sediment).

In Table 4 is shown values for r and p for elements in the Tisa sediments. From Table 3 it is clear that Al was singificantly positively correlated with Cu, Cr, Zn, Cd, and Ni.

Positive correlation existed among Al and Pb, but with less significance in the Pančevo alluvial sediments. The value of correlation coefficients (r) that are not at a significant level for Pb indicate that only a low percentage of Pb natural variability can be attributed to the textural and mineralogical variability of the sediments. Ti was significantly positively correlated with Cr, Li with Cu, Cr, and Ni, and Si was correlated with Cu.

Table 3. Pearson correlation matrix of elements in the Danube alluvial sediment

	Cu	Cr	Zn	Pb	Cd	Ni
Al	0.614*	0.732**	0.508**	0.318	0.465**	0.773**
Ti		0.362*				
Li	0.596**	0.628**				0.612**
Si	0.424*					

* Correlation is significant at the 0.05 level.

** Correlation is significant at the 0.01 level.

Table 4. Pearson correlation matrix of elements in the Tisa sediment

	Cu	Cr	Zn	Pb	Cd	Ni
Al	0.404*		0.503**	0.397*	0.434*	0.480**
Ti	-0.391*	-0.439*	-0.485*	-0.465*		
Li						0.775**
Si						

* Correlation is significant at the 0.05 level.

** Correlation is significant at the 0.01 level.

According to results from Table 4 it is clear that Al is positively correlated Cu, Zn, Pb, Cd, and Ni, while the Ni is correlated with Li in the Tisa sediments.

Ti content was negatively correlated with Cu, Cr, Zn, and Pb. Li is more strongly correlated with Ni then Al, but not positively correlated with other toxic elements and cannot explain at all the natural fluctuations of the content of other elements.

Therefore, in this paper as element for normalization is used Al, since this element is better suited to explain the natural variations in the concentrations most of elements in both types of sediment.

The scatter plots and the regression curves of Al and other investigated elements are presented, respectively, in Figure 2 (Tisa) and Figure 3 (Pančevo sediments), for the whole of the study area. Each diagram shows the regression line of investigated element on Al with the 95% confidence band.

Looking at Figure 2 (Tisa sediment) is possible to observe: contamination at sites 3, 4, 5, 6, 8, 9, 29 and 30 (Cu); 3, 4, 5, 6, 9, 11, and 29 (Cr); 3, 4, 5, 6, 11 , 29 and 30 (Zn); 5, 6, 9, 11 and 29 (Cd); 3, 4, 5, 6, 7, 8, 9, 10 and 11 (Ni) and 3, 4, 5, 6, 8, 9, 11 and 29 (Pb). Looking at Figure 3 (Pančevo sediment) is possible to observe there is contamination at sites 3 (strong) and 20 (with Cu), 21, 23, 25, 26, 29, 30, and 31 (with Cr), 10 (strong) and 7 (with Zn), 14, 20, 21, 22 and 24 (with Cd), 3, 13, 26, and 31 (with Ni) and 1 and 2 (with Pb).

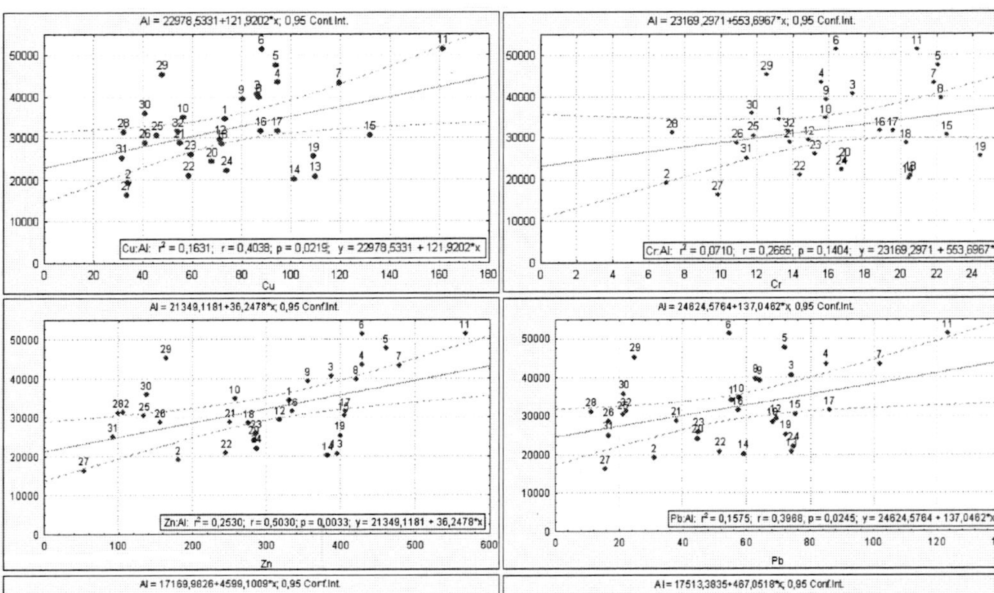

Figure 2. Al:element scatter plot for the Tisa and tributaries samples. Solid line represents the regression line; dashed lines define the 95% confidence band.

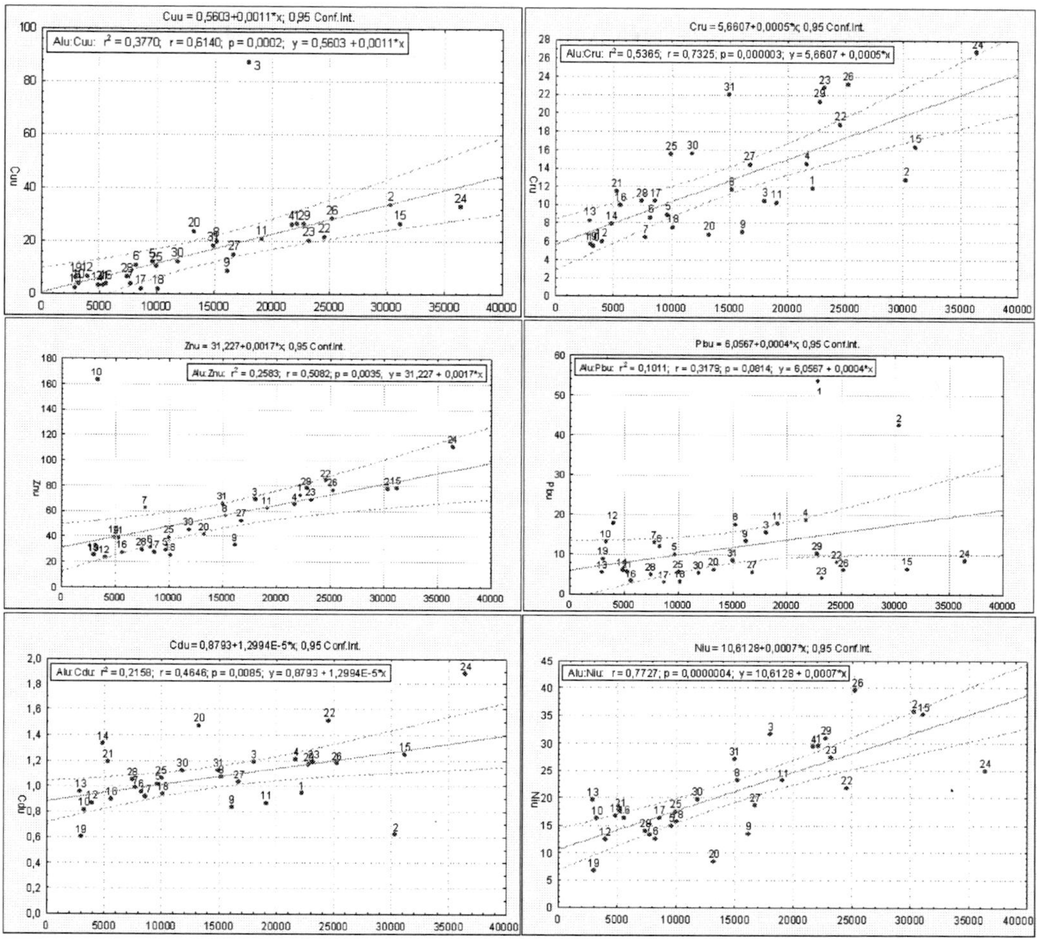

Figure 3. Al:element scatter plot for the Danube alluvial samples (Pančevo). Solid line represents the regression line; dashed lines define the 95% confidence band.

3.3. Comparison of Element Content with Background Values for Sediments and Soils

It is possible to observe the higher content of Cd, Cu, Pb, Ni and Zn in sediments of the Tisa in relation to sediment from Pančevo (Table 5).

Table 5. Statistical results for investigated sediments

Element	Min	Max	Mean	SD	Min	Max	Mean	SD
	Pančevo sediment				Tisa sediment			
Cd	0.60	1.88	1.07	0.26	2.04	4.97	3.25	0.89
Cr	5.52	26.71	12.50	5.85	6.99	24.44	16.14	4.55
Cu	1.72	87.10	16.76	16.51	31.61	162	74.87	31.30
Pb	2.99	54.40	11.65	11.01	11.17	123	54.60	27.36
Ni	6.67	39.50	20.90	8.33	17.28	54.58	31.25	9.70
Zn	23.53	164	55.78	30.23	53.94	567	297	131

Table 6. Comparison between background values for elements given by other authors and calculated for the alluvial and river sediment

Element	Ría de Vigo[a]	Continental Crust[b]	Ría de Vigo[c]	Tisa River (Serbia)[d]	Pančevo alluvial sediment[2]	River sediments[f]
Cd	-	0.1	-	-	0.95	0.003-1.967
Cr	43	126	34.04	11.37	8.62	13.80-86.22
Cu	25	25	29.41	41.97	10.77	16.84-90.63
Pb	25	14.8	51.29	19.3	11.97	16.87-82.997
Ni	30	56	30.32	42.27	12.48	28.56-51.4
Zn	100	65	105.35	127.3	31.45	74.86-325.5

[a] Barreiro (Ría de Vigo sediment) [20].
[b] Wedepohl (Continental crust) [21].
[c] Rubio et al. (Ría de Vigo sediment) [22].
[d] Sakan (Tisa river sediment) [4].
[e] Sakan (Pančevo alluvial sediment) [8].
[f] Sakan (river sediment in Serbia) [17].

The Ni content is not much higher compared to the background content (Table 6) for the Tisa and Danube alluvial sediment. The content of other elements (Cd, Cu, Pb and Zn) is much larger than the background values, and it is possible to conclude that there are significant anthropogenic sources of these elements in the basin of the Tisa River. Contamination with Zn, Cd, Pb and Cu are highly influenced by permanent pollution of Tisa river from agriculture, chemical factories, municipal sewage discharge and from mining industry and may be influenced by the input of these elements from the distant parts of the Tisa river from the countries in the Tisa watershade. On content of these elements a significant impact has also temporary accidents in Romanian mines during 2000 that have released heavy metals into the river flow. Since the Tisa flows roughly along the Romanian border and enters Hungary, it has received higher concentrations of pollutants. Cr content is the same in both types of sediment, while somewhat higher in the sediment from Pančevo. The amount of extractable Cr is not much in relation to the background content, so that it can be concluded that there are no significant anthropogenic sources of this element in the investigated sites. The high values of standard deviation for some element (Cu, Pb, Zn) also suggest anthropogenic sources for these elements.

3.4. The Result of Principal Component Analysis (PCA)

The multivariate principal component analysis (PCA) technique was applied in this research to the collection of metal concentration data, in order to explore the grouping of metals according to their similarities (R mode) and to group the similar sampling sites (spatial variability, Q mode). Kaiser-Meyer-Olkin (KMO) and Bartlett's sphericity tests were performed to examine the suitability of the data for PCA [23]. KMO is a measure of sampling adequacy that indicates the proportion of variance that is common, i.e., variance that may be caused by underlying factors. A high value (close to 1) generally indicates that principal component analysis may be useful, as was the case in this study, where KMO = 0.70 (Tisa sediment) and KMO = 0.77 (Danube alluvial sediments, Pančevo). Bartlett's tests of sphericity indicates whether a correlation matrix is an identity matrix, which would indicate that variables are unrelated [23]. The significance level of 0 in this study (less than 0.05) indicated that were significant relationships among the variables. PCA of the normalized variables (data set) was performed to extract significant principal components (PCs) and to further reduce the contribution of variables with minor significance; these PCs were then subjected to Varimax rotation (raw) to generate varifactors (VFs).

3.4.1. PCA Analysis of Examined Elements in the Tisa Sediments

R Mode

The results of the PCA for elements in the Tisa sediments: eigenvalues for different principal components, percentage variance accounted, cumulative percentage variance, and component loadings (Varimax rotated) are given in Table 7.

Varimax rotation was performed to secure increased principal components of chemical/environmental significance [24]. Eigenvalues greater than 1 were taken as criterion for the extraction of the principal components required for explaining the sources of variances in the data.

According to the results of the initial eigenvalues, two principal components are considered, which account for over 85% of the total variance (Table 7, Figure 4a). The initial component matrix for Tisa sediment indicates that Cu, Zn, Pb, and Cd are associated, displaying high values in the first component (PC1, 63% of total variance), while Ni is isolated in the second component (PC2, 22% of total variance). The PC1 probably represents anthropogenic sources, since PC2 represents lithogenic sources.

Q Mode

Applying the Q mode of PCA, the sediments are grouped by location, whereby three groups of sediment samples were isolated (Figure 4b).

One group consists of sediment samples with labels 25-32, the second group sediments with labels 1, 3-12, 15-18, 20-24, a third group of samples 13, 14 and 19, and the sample was marked with 2 is separate, but close to the first group (samples 25-32). Since the PCA performed with normalized metal content, it is possible to assume that these groups are isolated based on the level of contamination at the examined sites.

Sediment samples with labels 25-32 represent samples of tributaries and pools.

Content of elements in these sediments is lower than in river sediment, because on this area are no significant anthropogenic sources of toxic elements.

Second group of sediments consists mainly of the Tisa River sediments in which the increased content of toxic elements in relation to the sediments of tributaries and pools. These results may indicate permanent pollution, originating from different sources, and also the influence of accidental water pollution incidents on the Tisa sediment contamination.

The third group consisting of sediments with labels 13, 14, and 19 are sediments with similar content of elements, and sediment 13 and 14 are sampled close to one another at a distance of 10 km, so there is geochemical similarity among sediments, as well as similarities in anthropogenic sources. Content of trace element in sample of river sediment denoted with 2 is similar with element contents in tributaries, and therefore this sample is close to the group of river sediments.

Table 7. Total variance explained and component matrixes (Tisa sediment)

C*	Initial Eigenvalues			Extraction Sums of Squared Loadings			Rotations Sums of Squared Loadings		
	Total	% of Variane	Cumulative %	Total	% of Variance	Cumulative %	Total	% of Variance	Cumulative %
1	3.163	63.269	63.269	3.163	63.269	63.269	3.159	63.184	63.184
2	1.100	21.995	85.263	1.100	21.995	85.263	1.104	22.079	85.263
3	0.602	12.043	97.307						
4	0.079	1.580	98.887						
5	0.056	1.113	100.00						

*Component.

Table 8. Total variance explained and component matrixes (Danube alluvial sediment)

C*	Initial Eigenvalues			Extraction Sums of Squared Loadings			Rotations Sums of Squared Loadings		
	Total	% of Variance	Cumulative %	Total	% of Variance	Cumulative%	Total	% of Variance	Cumulative %
1	3.481	58.020	58.020	3.481	58.020	58.020	3.470	57.828	57.828
2	1.192	19.867	77.887	1.192	19.867	77.887	1.204	20.059	77.887
3	0.734	12.237	90.125						
4	0.386	6.426	96.551						
5	0.108	1.798	98.349						
6	0.099	1.651	100.00						

*Component.

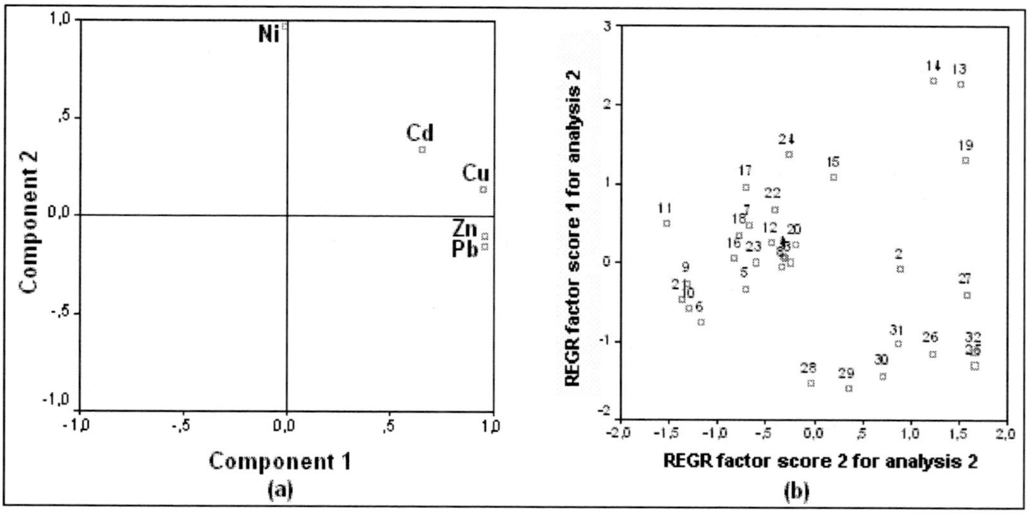

Figure 4. (a) Component Plot in Rotated Space. (b) Q mod of PCA, Tisa sediment.

3.4.2. PCA Analysis of Examined Elements in the Pančevo Alluvial Sediments

R Mode

The first component, which explains 58% of the total variance is related to the Cr, Zn, Pb, Cd, and Ni (Table 8). PC2 explains approximately 20% of the total variance and is positively loaded with Cu and Pb (Figure 5a).

In this case, PC1 and PC2 represent different anthropogenic sources. PC1 component is indicative of element origin from industrial complex at this location (refinery, Petrochemical Complex). PC2 is related to other contamination sources combined with traffic sources. Pb was correlated with the PC1 and PC2, which means that Pb comes from various anthropogenic sources.

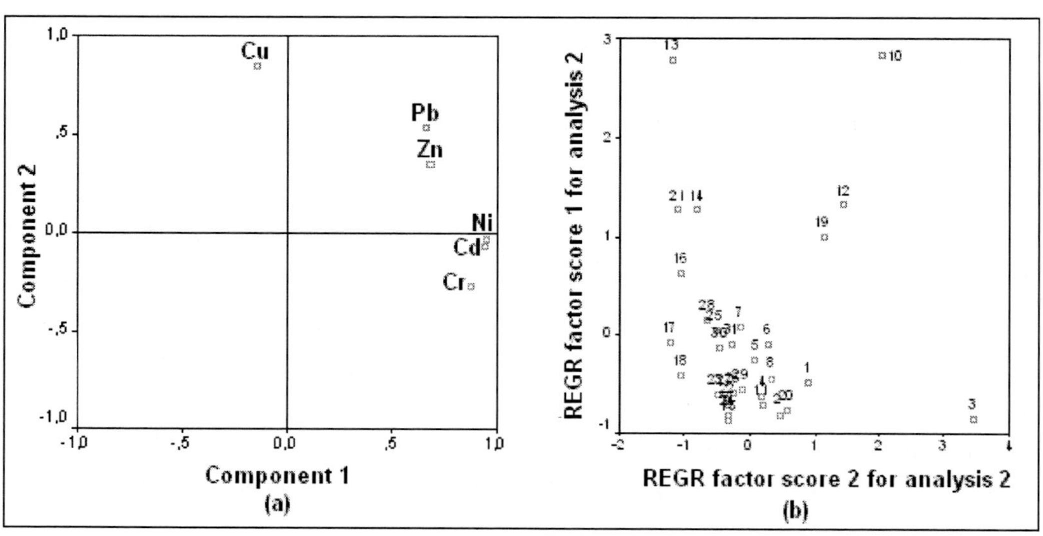

Figure 5. (a) Component Plot in Rotated Space. (b) Q mod of PCA, Pančevo sediment.

Q Mode

Most of the sediment samples constitute one group, where the samples with labels 13, 10, and 3 were separated from the other samples (Figure 5b). Sediment sample denoted with 10 contains high content of Zn in relation to others, while sample 3 is contain significantly more Cu. High content of Cu and Zn on these sites indicates the existence of local sources of contamination. Samples with labels 21 and 14, as well 12 and 19, form separate group that are closer to the most of the sediments from the investigated site.

CONCLUSION

Based on the above discussions, it may be concluded that geochemical normalization and statistical analysis are significant method for assessing trace element contamination and for distinguishing between natural and anthropogenic sources of trace elements in the sediments. In the case of the Tisa river sediments and Danube alluvial sediments from Pančevo, the following conclusions can be drawn: (1) Al is better suited to explain the natural variations in the concentrations most of elements in both types of sediment, and because of that, Al is used as element for normalization; (2) In the Tisa and Danube alluvial sediment, on some localities is observed high content of Cu, Cr, Zn, Cd, Ni, and Pb; (3) Higher content of Cd, Cu, Pb, Ni, and Zn in sediments of the Tisa in relation to sediment from Pančevo is observed; (4) The Ni content is not much higher compared to background content for the Tisa and Danube alluvial sediment. Cr content is the same in both types of sediment, while somewhat higher in the sediment from Pančevo. The amount of extractable Cr is not much in relation to the background content, so that it can be concluded that there are no significant anthropogenic sources of Ni and Cr on the investigated sites; (5) The content of other elements (Cd, Cu, Pb and Zn) is much larger than the background values, and it is possible to conclude that there are significant anthropogenic sources of these elements in the basin of the Tisa River. (6) The results of PCA (R mode) for Tisa sediment indicates that Cu, Zn, Pb, and Cd are associated, displaying high values in the first component, while Ni is isolated in the second component. The PC1 probably represents anthropogenic sources, since PC2 represents lithogenic sources. Applying the Q mode of PCA, the sediments are grouped by location, whereby three groups of sediment samples were isolated. These groups are isolated based on the level of contamination at the examined sites. (7) The first component in the PCA for Pančevo sediments is related to the Cr, Zn, Pb, Cd, and Ni is indicative of element origin from industrial. The PC2 is positively loaded with Cu and Pb and this component is related to other contamination sources combined with traffic sources. Results of Q mode of PCA indicate that most of the sediment samples constitute one group, where the samples with labels 13, 10, and 3 were separated from the other samples. High content of Cu and Zn on localities 3 and 10 in Pančevo sediments indicates the existence of local sources of contamination with these elements.

The results from this study show that investigated sediments are affected by combined trace elements contamination. Finally, the Pb, Zn, Cu, and Cd can be inferred to be tracers of anthropic pollution, whereas Ni and Cr were interpreted to be mainly inherited from parent materials.

ACKNOWLEDGMENTS

This research was supported by the Ministry of Science and Technological Development of the Republic of Serbia, Grant No. 172001.

REFERENCES

[1] Peñarroya, M. B. (2011). Trace element biogeochemistry in high mountain lake catchments: identifying anthropogenic versus natural components from the atmospheric contamination legacy in remote natural areas. *Doctoral Thesis*, University of Girona.

[2] Wang, X.S., Qin, Y. and Sang, S.X. (2005). Accumulation and sources of heavy metals in urban topsoils: a case study from the city of Xuzhou, China. *Environ. Geol.,* 48, 101-107.

[3] Sakan, S., Gržetić, I. and Đorđević, D. (2007). Distribution and Fractionation of Heavy Metals in the Tisza (Tisza) River Sediments. *Environ. Sci. Pollut. Res.,* 14, 229-236.

[4] Sakan, S., Đorđević, D., Manojlović, D. and Polić, P.(2009). Assessment of heavy metal pollutants accumulation in the Tisza river sediment. *J. Environ. Manage.,* 90, 3382-3390.

[5] Teng, Y., Ni, S., Wang, J. And Niu, L. (2009). Geochemical baseline of trace elements in the sediment in Dexing area, South China. *Environ. Geol.,* 57, 1649-1660.

[6] Sakan, S. and Đorđević, D. (2010). Evaluation of heavy metal contamination in sediments using the method of total digestion and determination of the binding forms-Tisa River Basin, Serbia, *J. Environ. Sci. Heal. A,* 45, 783-794.

[7] Sakan, S., Đorđević, D. and Manojlović, D. (2010). Trace elements as tracers of environmental pollution in the canal sediments (alluvial formation of the Danube River, Serbia), *Environ. Monit. Assess.,* 167, 219-233.

[8] Sakan, S., Đorđević, D. and Trifunović, S. (2011). Geochemical and statistical methods in the evaluation of trace element contamination: an application on canal sediments, *Polish J. of Environ. Stud.,* 20, 187-199.

[9] Relić, D., Đorđević, D., Sakan, S., Anđelković, I., Miletić, S. and Đuričić, J. (2011). Aqua regia extracted metals in sediments from the industrial area and surroundings of Pančevo, Serbia. *J. Hazard. Mater.,* 186, 1893–1901.

[10] Shikazono, N., Tatewaki, K., Mohiuddin, K.M., Nakano, T. and Zakir, H.M. (2012). Sources, spatial variation, and speciation of heavy metals in sediments of the Tamagawa River in Central Japan. *Environ. Geochem. Health.,* 34, 13-26.

[11] de Andrade Passos, E., Alves, J.C., dos Santos, I.S., Alves, J.P.H., Garcia C.A.B. and Costa, C.S. (2010). Assessment of trace metals contamination in estuarine sediments using a sequential extraction technique and principal component analysis. *Microchem. J.,* 96, 50-57

[12] Facchinelli, A., Sacchi, E. and Mallen, L. (2001). Multivariate statistical and GIS-based approach to identify heavy metal sources in soils. *Environ. Pollut.,* 114, 313-324.

[13] Bowen, H.J.M. (1979). Environmental Chemistry of the Elements; Academic Press: London, 1979.

[14] Nriagu, J. O., and J. M. Pacyna (1988). Quantitative assessment of worldwide contamination of air, water and soils by trace metals, *Nature*, 333, 134– 139.

[15] Szava-Kovats, R.C. (2008). Grain-size normalization as tool to assess contamination in marine sediments: Is the <63 μm fraction fine enough? *Mar. Pollut. Bullet.*, 56, 629-632.

[16] Aloupi, M. and Angelidis, M.O. (2001). Geochemistry of natural and anthropogenic metals in the coastal sediments of the island of Lesvos, Aegean Sea. *Environ. Pollut.,* 113, 211–219.

[17] Sakan, S., Đorđević, D., Dević, G., Relić, D., Anđelković, I. and Đuričić, J. (2011). A study of trace element contamination in river sediment in Serbia using microwave-assisted aqua regia digestion and multivariate statistical analysis, *Microchem. J.,* 99, 492-502.

[18] Reimann, C. and Filmozer, P.(1999). Normal and lognormal data distribution in geochemistry: death of a myth. Consequences for the statistical treatment of geochemical and environmental data. *Environ. Geol.,* 39(9), 1001–1014.

[19] Relić, D., Đorđević, D., Popović, A. and Blagojević, T. (2005). Speciations of trace metals in the Danube alluvial sediments within an oil refinery. *Environ. Internat.,* 31, 661-669.

[20] Barreiro, R. (1991) Estudio de metales pesados en medio y organismos de un ecosistema de Rõa (Pontedeume, A Coru~na). Tesis Doctoral, Universidad de Santiago de Compostela.

[21] Wedepohl, K.H. (1995). The composition of the continental crust. Geochim. Cosmochim. Acta, 59 (7), 1217-1232.

[22] Rubio, B., Nombela, M.A. and Vilas, F., (2000). Geochemistry of major and trace elements in sediments of the Ria de Vigo (NW Spain): an assessment of metal pollution. *Mar.* Pollut. Bullet., 40 (11), 968–980.

[23] Varol, M. (2011). Assessment of heavy metal contamination in sediments of the Tigris River (Turkey) using pollution indices and multivariate statistical techniques. *J. Hazard. Mater.,*195, 355-364.

[24] Kuppusamy, M.R. and Giridha, V.V. (2006). Factor analysis of water quality characteristics including trace metal speciation in the coastal environmental system of Chennai Ennore. *Environ. Internat.,* 32, 174-179.

In: Trace Elements ISBN: 978-1-62081-376-8
Editors: Diego A. De Leon and Paloma Raquel Aragon

Chapter 7

ZINC AVAILABILITY TO FORAGE CROPS IN SOILS OF THE PAMPAS REGION, ARGENTINA

Silvana Irene Torri[1], Alejo Perez-Carrera[2] and Alicia Fernández-Cirelli[2]
[1]Cátedra de Química General e Inorgánica, College of Agronomy,
University of Buenos Aires, Argentina
[2]Centro de Estudios Transdisciplinarios del Agua,
Facultad de Ciencias Veterinarias, University of Buenos Aires, Argentina

ABSTRACT

Zinc (Zn) deficiencies occur all over the world. Due to the naturally high soil fertility of the Pampas region, Argentina, macro and micronutrients have not been applied to soils through fertilization in a massive way. As a consequence, some trace elements deficiencies were noticed in some forage crops during the last years as a result of nutrient depletion. Trace mineral deficiencies were detected in dairy cows, causing a variety of suboptimal performances such as poor production and reproductive inefficiencies. Zinc and copper are the trace elements present in largest amounts in Buenos Aires City´s biosolids. Soils amended with biosolids have been reported to have high Zn availability. Therefore, biosolids land application may raise the concentration of Zn in forage crops, reducing the need for mineral fertilisers or supplements in the diet of dairy farms. This chapter address sound soil and crop management practices that may not only enhance crop productivity but also help in reducing Zn deficiencies in dairy cattle.

INTRODUCTION

Trace elements deficiencies in crops occur all over the world, but variations in soil conditions, climate, crop genotypes and management result in marked variations in the incidence of these problems. Some trace elements are essential for human beings, like iron (Fe), zinc (Zn), selenium (Se), iodine (I) and cobalt (Co). Among these, Fe, Zn and I are known to be limiting in the diets of a vast number of people (Kennedy et al., 2003; WHO, 2002).

The importance of certain trace elements for optimum growth and development of plants has been reported for the last century (Barker and Pilbeam, 2007; Cakmak, 2002; Marschner, 1995). Several trace elements are known to be essential for certain plants. Some others have been shown necessary only for a few species, whereas others are known to have stimulating effects on plant growth, although their functions are not yet recognized (Barker and Pilbeam, 2007). In Argentina, some essential trace elements deficiencies were noticed in field crops during the last years as a result of high-yield genotype crops associated with greater nutrient removal, soil degradation, and the adoption of improved agricultural mechanization (Andrade and Sadras, 2000; Urricarriet and Lavado, 1999).

Trace element deficiencies in cattle usually arise as a consequence of its inherent deficiency in soils producing forage, consumed through grazing or conserved as silage. The absence or deficiency of certain trace elements in animals can result in ill-thrift, reproductive failure or reduction in production (Crawshaw, Caldow 2005). The latter commonly occurs when trace elements deficiency corresponds to the phases of growth, reproduction, or lactation. Because of environmental, nutrient, disease, genetic, or other interactions, deficiencies of single or multiple trace elements may even occur when trace elements are being supplied at recommendated levels.

Zinc (Zn) is an essential trace element in plant and animal nutrition, and the second most abundant transition metal in organisms after Fe. It plays an essential role as a component of the six enzyme classes (Enzyme Commission number, EC 1–6), which include oxidoreductases, transferases, hydrolases, lyases, isomerases and ligases (Webb, 1992). The essentiality of Zn in plants was first shown in maize, and subsequently in barley and dwarf sunflower (Mazé, 1915, Sommer, Lipman 1926). Unlike other micronutrients, Zn deficiency is ubiquitous: it occurs in cold and warm climates, in drained and flooded soils and in acid and alkaline soils. In most crops, the typical leaf Zn concentration required for adequate growth approximates 15–20 mg Zn kg^{-1} DW (Marschner, 1995). In animals, Zn fulfills many biochemical functions in metabolism: it is an indispensable component of a number of enzymes or other proteins in animals (Hambidge et al, 1987; Miller, 1991; Gupta et al, 2008) and an essential element for maintaining normal growth, reproduction, and lactation performance (Graham, 1991; Gupta et al, 2008).

Ruminants without mineral supplements meet their Zn requirements from grazing pastures/herbage and soil ingestion. The recommended dietary allowance for zinc is 40 mg kg $^{-1}$ DM. Animals fed diets containing less than 5 mg Zn kg^{-1} develop signs of severe Zn deficiency, which include loss of appetite, reduced growth and inmunocompetence, reduced efficiency of conversion of feed energy to milk or live-weight gains, loss of hair and wool and keratotic skin lesions (Underwood, 1981). Other symptoms of zinc deficiency in ruminants are inflamed membranes of the nose and mouth. Zn deficiency causes parakeratosis, which is manifested by slow growth, skin lesions and hair loss. Cattle can develop spots on the tender parts of the skin, not unlike those occurring during photosensitization. Ruminants can tolerate high levels of Zn, although feeding it in excess - i.e.500 mg kg^{-1} DM - can produce toxicity. Excessive dietary supplements of zinc can result in sluggishness, loss of appetite and diarrhoea.

Argentina is the fifth largest producer of cattle in the world, contributing 2.5 million tonnes or 4.7% of world production. Two different farming regimes are used in Argentina for beef production: grass pasture and feedlot-based farming. Natural grasslands are the main sources of feeding for extensive farming practices in the Pampas region, Argentina (Cauhépé and Hidalgo, 2005). Animals grazing through the whole year produce meat with less cholesterol than through feedlot (García et al., 1996). Pasture forages are expected to satisfy the major part of the animal's nutrient requirements for energy, protein, vitamins, and minerals. Perennial ryegrass is the preferred forage grass in temperate regions of the world because of its high palatability and digestibility. However, plant nutrient requirements and composition do not always synchronize well with cattle nutrient needs.

The objectives of this chapter are to address sound soil and crop management practices that may not only enhance crop productivity but also help in reducing Zn deficiencies in dairy cattle.

Characteristics of the Study Area

The Pampas Region, Argentina, is one of the largest temperate field cropland areas of the Southern Hemisphere. It is located between 32° to 39°S and 56 to 67°W (Figure 1). This zone covers more 52 Mha of agriculturally prime quality land. The region also includes some areas which are marginal or unsuitable for cropping and are devoted to husbandry. The climate is humid, characterized by long warm summers and mild winters. Mean annual rainfall ranges from 600 mm in the west to 1200 mm in the east, whereas mean annual temperature ranges from 14 °C in the south to 21 °C in the north. Predominant soils are Mollisols formed on loess like materials of eolian origin, and to a lesser degree from fluvial sediments (Imbellone et al., 2010). The humid and semiarid subregions are characterized by Udolls and Ustolls, respectively, with minor occurrence of Aquolls in wet flat areas (Moscatelli and Pazos, 2000). These soils have naturally adequate concentration of trace elements and show no signs of contamination with potentially trace elements (Lavado et al., 1998; 2004).

The Pampas region maintained its native grasslands until the last quarter of 19th century and the beginning of the 20th century, when the establishment of perennial pastures and annual crops by European immigrants started (Viglizzo et al., 2001).

Extensive agriculture and cattle production are dominant activities in the region, playing a fundamental role in the economy of Argentina (Deybe and Flichman 1991) and in global food production (Imhoff et al. 2004). The Argentine Pampean Region has undergone major soil-use and management changes in recent years, moving away from mixed crop and livestock production systems to agricultural use or intensive livestock production systems.

The Pampean Region is the most important cattle production region in Argentina, accounting for approximately 60% of total cattle for beef and milk production (Rodriguez-Vazquez, 2010). In some areas where important cattle grazing systems are established, soil deficiencies in copper and zinc have been reported (Torri et al, 2012 *in press*). In the last twenty years, feedlots system are in expansion in this region, mainly due to the continuous increase in the surface used for soya bean crops (INDEC, 2002; SENASA, 2010). These cattle production systems use feed which incorporate core minerals whose basic components are Fe, Se, I, Mn, Co, Cu, and Zn. Trace element content in manure of intensive animal production systems is related to feed mineral content and the animal conversion efficiency (Nicholson et al., 1999).

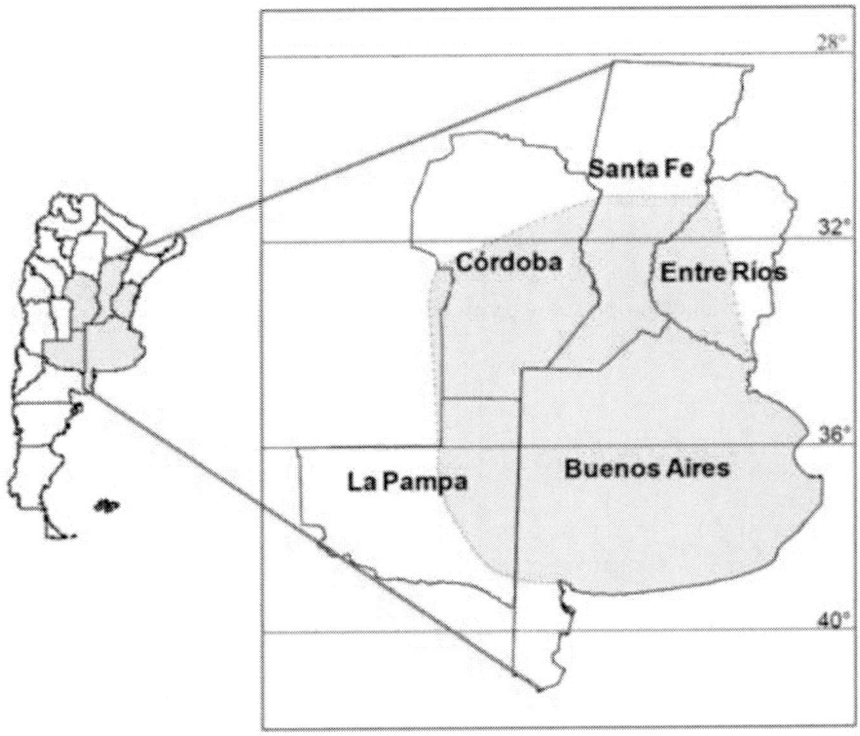

Figure 1. Map of the Pampas Region, Argentina.

Zn Availability in Soils of the Pampas Region

The total Zn content of the Pampas soil are within the values find in normal non contaminated soils of the world (Lavado et al., 2004). The distribution of total Zn between soil fractions is mainly determined by soil-specific precipitation, complexation and adsorption reactions. Soil type, pH, soil moisture, mineral and clay types and contents, diffusion and mass flow rates, weathering rates, soil organic matter, soil biota and plant uptake will also affect Zn distribution.

Total Zn content in soils usually includes fractions that are not immediately available to plants, micro-organisms and soil fauna (Zhang et al., 2001). From the agricultural point of view, the availability of trace elements is more important than their total concentration. Due to the naturally high soil fertility of the Pampas region, macro and micronutrients have not been applied in a massive way to soils through fertilization.

The lack of nutrients reposition resulted, with time, in a depletion of soil phytoavailable micronutrients (Urricariet and Lavado 1999). A wide variety of extractants have been used for estimating the availability of Zn in soils, with variable degrees of success. Among chelates, diethylenetriaminepentaacetic acid (DTPA) and ethylenediaminetetraacetic acid (EDTA) have been considered the most suitable extracting agents (Moral et al., 2002; Quevauviller et al., 1998). Figure 2 shows the effect of years of agriculture on DTPA available Zn on pristine soils in the in the centre of Santa Fe province (C. Miretti, *personal communication*).

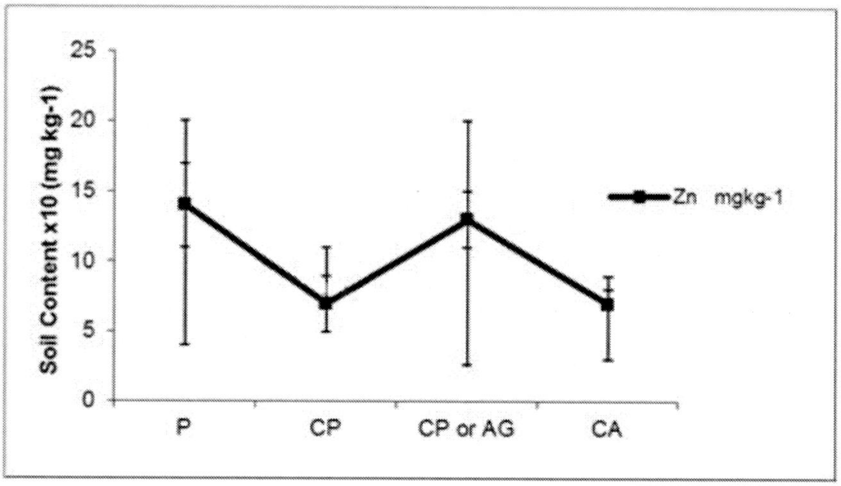

Figure 2. The effect of agriculture on DTPA extractable B and Zn. P: Pristine soils; CP: cattle production grazing pastures all year round in the field, as usual in the region; CP or AG: alternate animal grazing and agriculture and CA: continuous agriculture (C. Miretti, *personal communication*).

Table 1. Zinc availability in representative soils of the Pampas region

Extractant	Soils	Zone	Range (mg Zn kg^{-1})	References
EDTA	Argiudolls	N Buenos Aires	2.1-13.2	Urricariet and Lavado (1999)
	Haplustolls	La Pampa	1.4-15.0	Gonzalez and, Buschiazzo (1998)
DTPA	Argiudolls	N Buenos Aires	1.47	Lavado and Porcelli (2000)
	Argiudolls	NW Buenos Aires and S Santa Fé	1.1-5.4	Ratto de Miguez and Fatta (1990)
	Typic Argiudoll	N Buenos Aires	0.30-2.60	Ratto et al. (1997)
	Argiudolls and Haplustolls	S Córdoba	0.40-2.50	Buffa and Ratto (2005)

Marked variations in the range of Zn availability in soils of the Pampas region have been observed (Table 1).

In addition to single extractions, sequential extraction is a useful technique for determining the nature of the association of trace elements with soil components. Such information is potentially valuable for predicting bioavailability, leaching rates and transformations between physicochemical forms in agricultural and polluted soils (Torri and Lavado 2008 a,b). Torri and Lavado (2008 b) studied the distribution of Zn among different fractions in the superficial horizon of three typical pristine soils of the Pampas region using the sequential chemical extraction procedure described by McGrath and Cegarra (1992). Native Zn was mainly found as RES-Zn, exceeding 84% of the total content (Table 2). The order of Zn concentration in the fractions of the three pristine soils was: RES-Zn > INOR-Zn > OM- Zn > EXCH-Zn (Table 3). This result is typical for Zn found in non-contaminated soils (Shuman, 1999) where it is mostly found in unreactive forms in the crystal lattices of the minerals.

Table 2. Soil distribution of endogenous Zn in the three pristine soils of the Pampas region, among water-soluble and exchangeable fraction (EXCH), organic matter bound fraction (OM), inorganic precipitate fraction (INOR) and residual fraction (RES) (adapted from Torri and Lavado (2008 b)

Soils	EXCH-Zn	OM- Zn	INOR-Zn	RES-Zn
	mg kg $^{-1}$			
Typic Hapludoll	1.61 ±0.07	2.62 ±0.28	3.90 ±0.40	46.87 ± 0.75
Typic Natraquoll	1.07±0.07	1.75 ±0.04	4.14 ±0.07	40.04 ± 0.11
Typic Argiudoll	1.00 ±0.03	1.97 ±0.06	6.22 ±0.26	49.80 ± 0.29

Levels of Zinc in Pastures of the Pampas Region

In extensive agriculture, dairy cattle are dependent upon forage nutrient status to meet their nutritional needs. However, plant nutrient requirements and composition do not always synchronize well with cattle nutrient needs. The desired range of zinc levels considered adequate in pasture and animal feeds is 30 to 50 mg kg^{-1}.

There are few available reports on Zn concentration in forage species in the pampas region. Of all forage species, *Lolium perenne* is one of the most studied. *L. perenne* is a cool-season perennial bunchgrass which is widely distributed throughout the world. It is the preferred forage grass in temperate regions of the world, including the Pampas Region, Argentina. High growth rates under high fertility and an extensive root system make *L. perenne* also valuable for use in nutrient recycling systems.

Tissue Zn concentrations in L. *perenne* ranged from 64.2 - 19.9 mg kg^{-1} (DW) when cultivated in different textured pristine soils of the Pampas region (Torri, Lavado 2009). Tissue Zn concentrations depends on Zn availability in soils. Significant differences in Zn tissue nutrient concentrations were observed between harvest. In a pot experiment using three pristine soils of the Pampas region, the concentration of Zn in aerial biomass was within the range of 64.24-27.51 mg kg^{-1} (DW) in the first harvest. In the third harvest, a significant decrease of Zn in aerial biomass concentration compared to the first harvest was observed, irrespective the soil considered. The decrease in Zn concentration was probably originated by an initial depletion of available Zn. The element concentration in the above ground tissue was within the range of 28.05-19.9 mg kg^{-1} (DW).

In the southeast of the province of Córdoba, Argentina, Pérez Carrera (2006) reported Zn soil levels in the range of 61 - 129,5 mg kg^{-1}, whereas in alfalfa (Medicago sativa) the range was 20.2 - 56.8 μg g^{-1}.

Use of Soil Amendments for Zn Fortification of Pastures

Micronutrient supplementation can be provided not just by inorganic fertilisers, but also by addition of various sources of organic matter (manures, sewage sludge, crop residues, etc.). Research (National Research Council NRC 1984, 2000) has illustrated the potential improvement of animal health with dietary fortification of trace minerals. Trace elements shown to be required in the diet tend to concentrate in certain tissues in direct proportion to the amount supplied in the diet.

BIOSOLIDS

In some countries, the use of biosolids as an organic amendment is encouraged because of its high organic-matter content and concentration of macro and micronutrients. Biosolids of Buenos Aires City have a high concentration of available Zn, and may be used as a soil amendment (Torri, 2001). Numerous studies suggest there is little potential for EPT to accumulate in forages from one to three years of biosolids applications, even at greater doses than usual agronomic rates. However, the use of this residue as an organic amendment must be carefully studied to ensure that grazing animals are not exposed to undesirable concentrations of trace elements, in order to protect the human food chain. The NRC (1980) has set the domestic animal mineral tolerance level for Cu at 100 mg Cu kg^{-1} feed (except for sheep, which is set at 25 mg Cu kg^{-1} feed) and for Zn at 300 mg Zn kg^{-1} feed (except for Japanese quail, which is set at 125 mg Zn kg^{-1} feed). uptake of Cd, Cu, Pb, and Zn in soils amended.

A pot experiment was set up to investigate the effects of two kinds of biosolids on Zn concentrations in the aereal biomass of *Lolium perenne* L. The pot experiment was conducted using three representative Mollisols (U.S. Soil Taxonomy) of the Pampas Region, Argentina.

The soils used are classified as Typic Hapludol, Typic Natraquol and Typic Argiudol, and were sampled near Carlos Casares (35° 37' S - 61° 22' W), Pila (36° 1' S - 58° 8' W) and San Antonio de Areco (34° 15' S - 59° 29' W) towns respectively. Composite soil samples (10 sub samples, 0-15 cm depth) were collected from fields with no previous history of fertilization or contamination.

Table 3. Main physical and chemical characteristics of the three untreated soils (A horizon, 0-15 cm) used for pot experiment

	Typic Hapludoll	Typic Natraquoll	Typic Argiudoll
Clay (%)	19.2	27.6	30.3
Silt (%)	23.2	43	53.6
Sand (%)	57.6	29.4	16.1
pH	5.12	6.21	5.72
Organic carbon (g kg^{-1})	28.6	35.31	24.5
Electrical conductivity (dS m^{-1})	0.61	1.18	0.7
Cation exchange capacity ($cmol_{(c)}$ kg^{-1})	20.3	22.3	24.5
Exchangeable cations			
Ca^{2+} ($cmol_c$ kg^{-1})	10.2	9.1	12.6
Mg^{2+} ($cmol_c$ kg^{-1})	2	5.4	4.3
Na^+ ($cmol_c$ kg^{-1})	0.3	2.1	0.2
K^+ ($cmol_c$ kg^{-1})	2.8	1.6	2.1

Table 4. Selected properties of biosolids (B) and 70:30 DMW mixture of biosolids and incinerated biosolids ash (AB)

	B	AB
pH	5.82	6.17
Moisture content (%)	5	4.5
Total organic carbon (mg g^{-1})	251	176
Total N (mg g^{-1})	19.3	22.5
Total P (mg g^{-1})	0.052	0.086
Electrical conductivity (dS m^{-1})	0.90	0.89
Cation exchange capacity ($cmol_c$ kg^{-1})	11.95	nd
Ca (mg g^{-1})	22.5	nd
Mg (mg g^{-1})	5.6	nd
K (mg g^{-1})	10.7	nd
Total Cd (mg kg^{-1})	10.08	13.08
Total Cu (mg kg^{-1})	750.8	894.7
Total Pb (mg kg^{-1})	334.2	365.9
Total Zn (mg kg^{-1})	2500	3200

nd = not determined.

Non-digested biosolids from Aldo Bonzi wastewater treatment plant located at the SW outskirts of Buenos Aires City. The biosolids (B) was dried at 60ºC before grinding and sieving (<2 mm) and then split into two portions. A portion was incinerated at 500º C. The ash obtained was thoroughly mixed with a portion of the sieved biosolids, resulting in a new mixed waste which contained 30% DM as ash (AB). Both amendments (B, AB) were thoroughly mixed with each of the three air-dried soils to simulate application rates of 150 Mg ha^{-1} in the field. Pots were left undisturbed and allowed to settle down over 60 days. Afterwards, *L perenne* seeds were sown and harvested 8, 12, 16 and 20 weeks after sowing, by cutting just above the soil surface. Aerial biomass was oven dried at 65°C for 48 h to reach constant dry weigh. Chemical properties of the soils, B and BA are shown in Tables 3 and 4.

The contents of Cd, Cu, Pb and Zn in SS and AS did not exceed ceiling concentrations for land application recommended by Argentine regulation (S.A.D.S, 2001).

Lolium Perenne L. Biomass and Zn Concentration

The germination of *L. perenne* in both sludge amended soils in the pot experiment was delayed for 15 days, possibly conditioned by the phytotoxic potential of B or BA (Zucconi et al, 1985). The intense mineralization of the labile organic matter pool of the sludge (Torri et al, 2003), which may have originated ammonia, low molecular weight organic acids and/or salts, all of which have been shown to have inhibitory effects (Wong et al, 1983; Chaney 1983; Adriano et al, 1973). Other studies have also reported that this toxic effect disappeared within 14 to 21 days (Benítez et al, 2001).

After emergence, *L. perenne* grew uniform in both sludge treatments along the growing period, showing no visible symptoms of metal toxicity or nutrient imbalances. Partial and total dry matter yields of *L perenne* grown in each treatment in the three soils are shown in Table 5.

Table 5. Partial and total mean values and standard deviation of aerial dry weight (g) of *L. perenne* grown in control and sludge-treated pots over four harvests (n = 3, ±S.E.). Soils: Typic Hapludoll (H), Argiudoll (A) and Natraquoll (N). Treatments: C= control, B= biosolids amended soils, AB= soils amended with the 70:30 DMW mixture of biosolids and incinerated biosolids ash. Groups in a column detected as different at the 0.05 probability level (Tukey test) were marked with different letters (a, b, c, etc. for partial harvest; A,B,C, etc. for total yield)

	1° harvest		2° harvest		3° harvest		4° harvest		Total DM yield	
	aerial dry weight (in g) of *L. perenne* / pot									
H - C	2,43 ± 0,05	ab	2,66 ± 0,11	ab	3,01 ± 0,13	bc	1,73 ± 0,10	b	9,83	BC
H - B	2,45 ± 0,07	ab	3,01 ± 0,10	a	5,49 ± 0,14	a	4,38 ± 0,30	a	15,32	A
H - AB	2,82 ± 0,10	ab	2,99 ± 0,06	a	5,82 ± 0,10	a	3,14 ± 0,07	ab	14,78	A
N - C	1,97 ± 0,09	b	1,76 ± 0,13	b	1,11 ± 0,13	c	0,84 ± 0,06	c	5,68	C
N - B	2,22 ± 0,13	ab	2,45 ± 0,13	ab	4,99 ± 0,23	ab	4,65 ± 0,25	a	14,32	A
N - AB	2,32 ± 0,12	ab	2,35 ± 0,09	ab	4,31 ± 0,31	ab	4,17 ± 0,44	a	13,15	AB
A - C	2,51 ± 0,16	ab	1,96 ± 0,16	b	1,58 ± 0,16	c	0,78 ± 0,02	c	6,82	C
A - B	3,07 ± 0,07	a	3,28 ± 0,02	a	5,28 ± 0,35	a	4,50 ± 0,26	a	16,13	A
A - AB	2,87 ± 0,13	ab	2,96 ± 0,05	a	4,74 ± 0,22	ab	3,33 ± 0,23	ab	13,90	AB

Both sludge amendments resulted in an increase in plant aerial biomass during the experimental period compared with plants grown in unamended soils, with no significant differences between AB and B treatments. Several factors may have contributed to improve growth in both sludge amended soils, especially the increased supply of N and P, in agreement with Antolin et al (2005) and Hseu, Huang (2005), together with an increasing limitation on nutrient supply in control soils with time.

In the first harvest, a significant increase of Zn in aerial biomass concentration in amended soils compared to controls was observed, irrespective the soil considered. No significant differences ($p<0.05$) in aerial biomass concentration of Zn between B or AB treatments were observed for the same soil.

The highest Zn concentration was detected in *L.perenne* grown on the Hapludol amended with biosolids (378.9 mg Zn kg^{-1} DMW), a significantly higher concentration compared to the Argiudol amended with biosolids (177.6 mg Zn kg^{-1} DMW). In the AB treatments, no significant differences among soils were observed. In the third harvest, a significant decrease of Zn in aerial biomass concentration compared to the first harvest was observed in both biosolids amended soils, irrespective the soil considered.

However, plants sown in the three amended soils did not show significant differences in terms of Zn concentration in aerial tissue, regardless soil or biosolids amendment (Figure 3).The decrease in Zn concentration was probably originated by an initial depletion of available biosolids-borne elements. Nevertheless, Zn concentrations in the above ground tissue of *L.perenne* grown in the amended soils were still significantly higher than controls, indicating high availability.

L. perenne grown in soils amended with AB not exhibit significantly higher Zn concentration in aerial biomass compared to B treatment in the three soils. These results are in agreement with previous studies, in which incineration was found to reduce the availability of Cu and Zn, increasing the percentage of residual fractions (Torri, Lavado 2008a). These findings are in good agreement with previous results reported by Obrador et al. (2001).

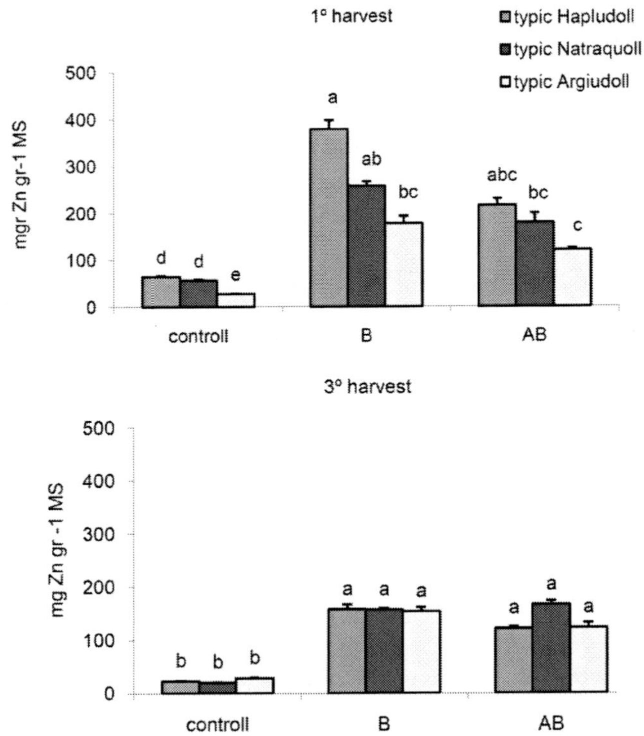

Figure 3. Accumulation of Zn (mg kg^{-1} DW) in shoots of *Lolium perenne* L. Soils: Typic Hapludoll, Argiudoll and Natraquoll. Treatments: C= control, B= biosolids amended soils, AB= soils amended with the 70:30 DMW mixture of biosolids and incinerated biosolids ash. For each harvest, groups detected as different at the 0.05 probability level (Tukey test) were marked with different letters.

Zinc concentration in shoots was in all cases below the range of critical concentration in plants described by Kabata-Pendias and Pendias (2001). Moreover, the concentration of these elements in aerial tissue was found to be under the threshold values specified by the NRC (1985) suggesting that consumption of *L.perenne* grown on biosolids amended soils would pose no risk to grazing animals.

This issue is crucial in order to avoid the threat of transfer of metals to the food chain. Physiological mechanisms that regulate the internal translocation of PTE have been postulated for this species (Santibáñez et al, 2008).

Animal Manure

Moscuzza et al (2011) analyzed the nutritional enrichment with Cu and Zn of *Lolium perenne* (ryegrass) through fertilization with bovine manure spiked with known concentrations of these trace elements. The experiment was conducted with agricultural soils from the province of Buenos Aires. Copper and Zn were applied to manure through increasing solution concentrations of sulfate salts: SN1, containing 7.5 mg of Cu and 60 mg of Zn; SN 2, containing 25 mg of Cu and 200 mg of Zn; and SN 3, containing 50 mg of Cu and 400 mg of Zn. Each 1 L solution was mixed by stirring with 1 kg of manure. These mixtures were incubated for 15 days at room temperature. Fertilization was performed at the surface or by homogenization through a mixed fraction of the soil with manure. The results showed differences in the concentrations of Cu and Zn between fertilized and unfertilized treatments. No differences in plant tissue concentration of Cu and Zn were observed between fertilization technology (homogenization or superficial).

REFERENCES

Adriano D., A. Chang, P. Pratt and R. Sharpless, 1973. Effect of soil application of dairy manure on germination and emergence of some selected crops, *J. Environ. Qual.* 2: 396–399.

Andrade, F y V Sadras. 2000. Bases para el manejo del maíz, el girasol y la soja. *Ed Médica Panamericana.* pp 207-232. Balcarce.

Antolin M.C., I. Pascaul, C. Garcia, A. Polo and M. Sanchez-Diaz. 2005. Growth, yield and solute content of barley in soils treated with sewage sludge under semiarid Mediterranean conditions, *Field Crops Res.* 94 , 224-237.

Barker A.V., Pilbeam D.J. 2007. *Handbook of Plant Nutrition*, Taylor and Francis Group Press, Boca Raton, FL.

Benitez, F.J., Acero, J.L., Real, F.J., Rubio, F.J., Leal, A.I., 2001. The role of hydroxyl radicals for the decomposition of p-hydroxy phenylacetic acid in aqueous solutions. *Water Res.* 35, 1338–1343.

Buffa, EV y SE Ratto. 2005. Disponibilidad de cinc, cobre, hierro y manganeso extraíble con DTPA en suelos de Córdoba (Argentina) y variables edáficas que la condicionan. *Ciencia del Suelo* 23:107-114.

Cakmak I. 2002. Plant nutrition research: Priorities tomeet human needs for food in sustainable ways, *Plant Soil* 247, 3–24.

Cauhépé M.A., Hidalgo L.G. 2005. La Pampa Inundable: el uso ganadero como base de la sustentabilidad social, económica y ambiental. [Flooding Pampa: the cattle use as base of the social, economic and environmental sustainable]. In Oesterheld M., Aguiar M., Ghersa Cl. And Paruelo J. (Eds.) La Heterogeneidad de la Vegetación de los Agroecosistemas. Un homenaje a Rolando León. *Editorial de la Facultad de Agronomía*, Universidad de Buenos Aires, Argentina. pp 401-412.

Chaney R.L. 1983. Potential effects of waste constituents on the food chain. In: J.F. Parr, Editor, *Land Treatment of Hazardous Wastes*, 50–76 (New Jersey).

Crawshaw, M. and Caldow, G. 2005. Trace element disorders in beef cattle. *Technical Note* 572. SAC.

Deybe, D., Flichman, G. 1991. A regional agricultural model using a plant growth simulation program as activities generator – an application to a region in Argentina. *Agricultural Systems* 37, 369–385.

García, P., Pensel, N., Margaría, C., 1996. *Grasa Intramuscular y Colesterol en carnes vacuna, de pollo y de pescado. Instituto Tecnologíaa de Carnes, CICV–INTA Castelar Boletín del Centro de Consignatarios Directos de Hacienda* 7 (92), 14–15.

Gonzalez, G y D Buschiazzo. 1998. Contenido de hierro, cobre, manganeso y cinc em suelos de la provincia de La Pampa, Argentina. *XV Congreso Argentino de la Ciencia del Suelo.*

Graham, T.W. 1991. Trace element deficiencies in cattle. *Vet Clin North Am Food Anim Pract.*; 7(1):153-215.

Gupta, U.; Kening, W. and Siyuan, L. 2008. Micronutrients in soils, crops and livestock. *Earth Science Frontiers*, 15 (5): 110-115.

Hambidge, K.M., Cassey, C.E. and Krebs, N.S. 1987. Zinc. In: Trace element in human and animal nutrition. Merts, W. (ed.). Academic Press, Orlando, 1-138.

Hseu Z.Y. and C.C. Huang. 2005. Nitrogen mineralization potentials in three tropical soils treated with biosolids, *Chemosphere* 59, 447-454.

Imbellone, P A, J E Giménez, JL Panigatti. 2010. Suelos de la Región Pampeana. *Procesos de formación. INTA*. 320 p.

Imhoff, M.L., L. Bounoua, T. Ricketts, C. Loucks, R. Harriss, et al. 2004. Global patterns in human consumption of net primary production. *Nature* 429:870-873

INDEC (National Institute of Statistics and Census). 2002. Agricultural National Census.Available in: *www.indec.gov.ar*

Kabatas-Pendias, A and H Pendias. 2001. *Trace elements in soils and plants*. Third edition. Boca Raton, FL: CRC Press.

Kennedy, G., Nantel, G., Shetty, P. 2003. The scourge of "hidden hunger": Global dimensionsof micronutrient deficiencies. *Food, Nutrition and Agriculture*, 32: 8-16.

Lavado, RS and CA Porcelli. 2000. Contents and main fractions of trace elements in Typic Argiudolls of the Argentinean Pampas. *Chemical Speciation and Bioavailability*. 12: 67-70.

Lavado, RS, M Rodríguez,J Scheiner,M Taboada, G Rubio, R Alvarez, M Alconada, and M Zubillaga. 1998. Heavy metals in soils of Argentina: comparison between urban and agricultural soils. *Commun. Soil Sci. Plant Anal.* 29: 1913–1917.

Lavado, RS, M Zubillaga, R Alvarez and MA Taboada. 2004. Baseline levels of potentially toxic elements in Pampas Soils. *Soil and Sediment Contamination* 13:427-437.

Marschner H. 1995 *Mineral Nutrition of Higher Plants, 2nd ed.*, Academic Press, London.

Mazé P. 1915. Détermination des éléments minéraux rares nécessaires audéveloppement du maïs. *Comptes Rendus Hebdomadaires des Séances de L'académie des Sciences* 160: 211–214.

McGrath, SP and J Cegarra. 1992. Chemical extractability of heavy metals during and after long-term applications of sewage sludge to soil. *Journal of Soil Science* 43: 313–321.

Miller, E.R., Lei, X and Ullrey, D. 1991. Trace elements in animal nutrition. In: Mortvedt, JJ, Cox, FR, Schuman LM, et al (eds.), *Micronutrients in Agriculture*, SSSA Book Series N°4, 2nd edition. Madison, 593-662.

Moral, R, RJ Gilke, and J Moreno-Casalles. 2002 A comparison of extractants for heavy metals in contaminated soils from Spain. *Commun. Soil Sci. Plant Anal.* 33:2781-2791.

Moscatelli, G and MS. Pazos. 2000. Soils of Argentina: Nature and Use. In: I. Kheoruenromne and S. Theerawong (Eds.). Bangkok, Thailand. *Proceedings of the International Symposium on Soil Science: Accomplishments and Changing Paradigm towards the 21st Century, and IUSS Extraordinary Council Meeting 81-92.*

Moscuzza, C., Pérez Carrera, A., Fernández Cirelli, A. 2011. Forage enrichment with micronutrients in beef grazing systems. *11th International Conference on the Biogeochemistry of Trace Elements (ICOBTE). Florence, Italy.*

Nicholson, F., B. Chambers, J. Williams and R. Unwin, 1999. Heavy metal contents of livestock feeds and animal manures in England and Wales. *Bioresource Technology*, 23, 23–31.

NRC (National Research Council). 1980. Mineral Tolerance of Domestic Animals. National Academy Sciences, Washington DC.

NRC (National Research Council). 1984. Nutrient Requirements of Beef Cattle. Sixth Revised Edition. Washington DC. National Academy Press.

NRC (National Research Council). 1985. Nutrient Requirements of Sheep. Sixth Revised Edition. Washington DC. National Academy Press.

NRC (National Research Council). 2000. Nutrient Requirements of Beef Cattle. Seventh Revised Edition. Washington DC. National Academy Press.

Obrador, A., Rico, M.I., Alvarez, J.M. and Novillo, J. 2001. Influence of termal treatment on sequential extration and leaching behaviour of trace metals in contaminated sewage sludge. *Biores. Technol.*, 76, 259-264.

Pérez Carrera, A. 2006. Trace elements in water, soil, forage and milk. PhD Thesis, Faculty of Veterinary Sciences, University of Buenos Aires.

Quevauviller, P, M Lachica, E Barahona, A Gomez, G Rauret, A Ure, and H Muntau. 1998. Certified reference material for the quality control of EDTA and DTPA-extractable trace metal contents in calcareous soil (CRM 600). Fresen. *J. Anal. Chem.* 360: 505-511.

Ratto de Míguez, S y N Fatta. 1990. Disponibilidad de micronutrientes en suelos del área maicera núcleo. *Ciencia del Suelo* 8: 9-15.

Ratto, SL, L Giuffre y C Sainato. 1997. Variación espacial de micronutrientes en suelo y planta en un Molisol. *Ciencia del Suelo* 15:39-41.

Rodriguez-Vazquez, G. 2010. Análisis comparativo de las existencias ganaderas 2008-2010. Informe Técnico N° 13.*Servicio Nacional de Sanidad y Calidad Agroalimentaria (SENASA), Argentina.*

Santibanez, C., Verdugo, C., Ginocchio, R., 2008. Phytostabilization of copper mine tailings with biosolids: implications for metal uptake and productivity of Lolium perenne. *Science of the Total Environment* 395, 1e10.

S.A.D.S. 2001. Secretaría de Ambiente y Desarollo Sustentable, Ministerio de Salud y Ambiente. *Resolución* 97/01, Anexo 1, pp.62.

SENASA (Servicio Nacional de Sanidad y Calidad Agroalimentaria). 2010. Estadísticas de Establecimientos de Engorde a Corral. [National Service of Health and Agrofood Quality. Statistics from feed lots]

Shuman, L.M. 1999. Organic waste amendments effect on zinc fractions of two soils. *J. Environ. Qual.*, 28, 1442-1447.

Sommer AL, Lipman CB. 1926. Evidence on the indispensable nature of zinc and boron for higher green plants.*Plant Physiology* 1: 231–249.

Torri S, Alvarez R, Lavado R. 2003. Mineralization of Carbon from Sewage sludge in three soils of the Argentine pampas. *Commun. Soil Sci. and Plant Anal.* 34 (13-14): 2035-2043.

Torri S, Lavado R. 2009. Plant absorption of trace elements in sludge amended soils and correlation with soil chemical speciation. *Journal of Hazardous Materials,* 166: 1459–1465. ISSN: 0304-3894 doi: 10.1016/ j.jhazmat.2008.12.075.

Torri S, Urricariet A.S, Lavado R. 2012. Micronutrient availability in crop soils of the Pampas region, Argentina. In: *Soil Nutrients*. Editor: M. Miransari, Nova Science Publishers, Inc., Hauppauge, NY 11788. In press, ISBN: 978-1-61324-785-3.

Torri SI. 2001. Distribución y biodisponibilidad de Cd, Cu, Pb y Zn en suelos fertilizados con biosolidos. *Tesis de Magister Scientiae en Ciencias del Suelo, FAUBA.*

Torri, S and R Lavado. 2008 a. Dynamics of Cd, Cu and Pb added to soil through different kinds of sewage sludge. *Waste Management*, 28: 821-832.

Torri, S and R Lavado. 2008 b. Zn distribution in soils amended with different kinds of sewage sludge. *Journal of Environmental Management*, 88: 1571-1579.

Underwood, E.J. (1981) *"The Mineral Nutrition of Livestock"*, 2nd ed. Commonwealth Agricultural Bureaux, Farnham Royal, Slough, England.

Urricariet, S. and R.S. Lavado. 1999. Indicadores de deterioro en suelos de la Pampa Ondulada. *Ciencia del Suelo* 17: 37-44.

Viglizzo, EF, FA Lertora, AJ Pordomingo, JN Bernardos, ZE Roberto and H Del Valle. 2001. Ecological lessons and applications from one century of low external input farming in the pampas of Argentina. *Agric Ecosyst Environ* 83:65–81.

Webb, E.C. 1992. *Enzyme nomenclature*. Academic Press, San Diego, CA.

Wong M.H., Y.H. Cheung and C.L. Cheung, 1983.The effects of ammonia and ethylene oxide in animal manure and sewage sludge on seed germination and root elongation of Brassica parachinensis, Environ. Pollut. A. 30: 109–123.

World Health Organization (2002). *The World Health Report* 2002. *Reducing Risks,* PromotingHealthy Life World Health Organization, Geneva.

Zhang, H, FJ Zhao, B Sun, W Davison, and SP McGrath. 2001. A new method to measure effective soil solution concentration predicts copper availability to plants. Environ. *Sci. Technol.* 35:2602—2607.

Zucconi, F., Monaco, A., Forte, M., De Bertoli, M., 1985. Phytotoxins during the stabilization of organic matter. En: Gasser, J.K.R (Ed.), *Composting of Agricultural and Other Wastes.* Elsevier, London, U.K, pp. 73-80.

INDEX

C

F

G

H

I

J

K

L

S